T0212395

Big Data Preprocessing

Julián Luengo • Diego García-Gil
Sergio Ramírez-Gallego • Salvador García
Francisco Herrera

Big Data Preprocessing

Enabling Smart Data

 Springer

Julián Luengo
Department of Computer Science and AI
University of Granada
Granada, Spain

Diego García-Gil
Department of Computer Science and AI
University of Granada
Granada, Spain

Sergio Ramírez-Gallego
DOCOMO Digital España
Madrid, Madrid, Spain

Salvador García
Department of Computer Science and AI
University of Granada
Granada, Spain

Francisco Herrera
Department of Computer Science and AI
University of Granada, Granada, Spain

ISBN 978-3-030-39107-2   ISBN 978-3-030-39105-8   (eBook)
https://doi.org/10.1007/978-3-030-39105-8

This Springer imprint is published by the registered company Springer Nature Switzerland AG.
The registered company address is: Gewerbestrasse 11, 6330 Cham, Switzerland

This book is dedicated to all the people with whom we have worked over the years and who have made it possible to reach this moment. Thanks to the members of the research institute "Andalusian Research Institute in Data Science and Computational Intelligence."
To our families.

Preface

The massive growth in the scale of data has been observed in recent years, being a key factor of the Big Data scenario. Big Data can be defined as high volume, velocity, and variety of data that require a new high-performance processing. Addressing Big Data is a challenging and time-demanding task that requires a large computational infrastructure to ensure successful data processing and analysis. Being a very common scenario in real-life applications, the interest of researchers and practitioners on the topic has grown significantly during these years. Among Big Data disciplines, data mining is a key topic, enabling the user to extract knowledge from enormous amounts of raw data. However, this raw data is not always in the best condition to be treated, analyzed, and surveyed. The application of preprocessing techniques is a must in real-world applications, to ensure quality data, Smart Data, for a proper treatment and analysis. The term Smart Data refers to the challenge of transforming raw data into quality data that can be appropriately exploited to obtain valuable insights.

This book aims at offering a general and comprehensible overview of data preprocessing in Big Data, enabling Smart Data. It contains a comprehensive description of the topic and focuses on its main features and the most relevant proposed solutions. Additionally, it considers the different scenarios in Big Data for which the application of data preprocessing techniques can suppose a real challenge. Data preprocessing is a multifaceted discipline that includes data preparation, compounded by integration, cleaning, normalization, and transformation of data; data reduction tasks such as feature selection, instance selection, and discretization; and resampling techniques to deal with imbalanced data.

This book stresses the gap with standard data preprocessing techniques and their Big Data equivalents, showing the challenging difficulties in their development for the latter. It also covers the different approaches that have been traditionally applied and the latest proposals in Big Data preprocessing. Specifically, it reviews data reduction methods, imperfect data approaches, discretization techniques, and imbalanced data preprocessing solutions. Finally, this book describes the most popular Big Data libraries for machine learning, focusing on their data preprocessing algorithms and utilities.

Granada, Spain   Julián Luengo
Granada, Spain   Diego García-Gil
Madrid, Spain   Sergio Ramírez-Gallego
Granada, Spain   Salvador García
Granada, Spain   Francisco Herrera
June 2019

Contents

1	**Introduction**		1
	1.1	Big Data	1
	1.2	Big Data Analytics	2
	1.3	Big Data Preprocessing	4
		1.3.1 Data Reduction	6
		1.3.2 Imperfect Data	9
		1.3.3 Imbalanced Datasets	10
	1.4	Big Data Streaming	11
	References		13
2	**Big Data: Technologies and Tools**		15
	2.1	Introduction	15
	2.2	Basic Concepts and Techniques	16
		2.2.1 Infrastructure	16
		2.2.2 Storage and Access	17
		2.2.3 Distributed Processing Engines	19
		2.2.4 Other Service Management Components	22
	2.3	Hadoop Ecosystem	23
		2.3.1 Distributed Storage Components	24
		2.3.2 Distributed Processing Components	26
		2.3.3 High-Level Components	26
	2.4	Apache Spark	29
		2.4.1 Spark Processing Engine	29
		2.4.2 MLlib	32
		2.4.3 Spark Packages	35
	2.5	Streaming Processing Frameworks for Big Data	37
		2.5.1 Apache Flink	37
		2.5.2 Apache Flume	39
		2.5.3 Apache Storm	39

2.6   Spark vs. Flink: A Thorough Comparison Between Two
      Outstanding Platforms 40
References 41

3   **Smart Data** 45
    3.1   Introduction 45
    3.2   From Big Data to Smart Data 46
    3.3   Smart Data and Internet of Things: Smart Cities and Beyond 48
    References 49

4   **Dimensionality Reduction for Big Data** 53
    4.1   Introduction 53
    4.2   The Curse of Big Dimensionality 54
    4.3   Distributed Proposals for Dimensionality Reduction 58
    4.4   An Information Theoretical Feature Selection Framework
          for Apache Spark 62
          4.4.1   Information Theory-Based Filter Methods 63
          4.4.2   Feature Selection Filtering Framework for Big Data 65
          4.4.3   High-Dimensional Feature Selection: An
                  Experimental Framework 69
    4.5   Dimensionality Reduction in Big Data Streaming 71
          4.5.1   Information Gain 71
          4.5.2   OFS: Online Feature Selection 73
          4.5.3   FCBF: Fast Correlation-Based Filter 73
    4.6   Summary and Conclusions 76
    References 77

5   **Data Reduction for Big Data** 81
    5.1   Introduction 81
    5.2   Parallel and Rapid Prototype Reduction 83
    5.3   MRPR: A MapReduce Solution for Prototype Reduction
          in Big Data Classification 85
          5.3.1   Map Stage: Distributed Prototype Reduction 86
          5.3.2   Reduce Stage: Aggregation of Partial Results 87
          5.3.3   On the Election of Prototype Reduction Methods
                  for Distributed Systems 87
    5.4   Transforming Big Data into Smart Data Through Data
          Reduction 89
    5.5   Nearest Neighbor Classification for High-Speed Big Data
          Streams Using Spark 92
          5.5.1   Partitioning 93
          5.5.2   Updating the Distributed Tree with Edition 94
          5.5.3   Prediction 95
    5.6   Summary and Conclusions 96
    References 97

6   Imperfect Big Data 101
    6.1   Introduction 101
    6.2   Noise Filtering 102
    6.3   Enabling Smart Data: Noise Filtering in Big Data Classification ... 104
        6.3.1   HME-BD: Homogeneous Ensemble 105
        6.3.2   HTE-BD: Heterogeneous Ensemble......................... 106
        6.3.3   ENN-BD: Similarity Based Method......................... 108
        6.3.4   Noise Filtering: An Experimental Study 109
    6.4   Noisy Big Data Treatment with the KNN Algorithm 113
    6.5   Missing Values Imputation 114
    6.6   Summary and Conclusions 117
    References.. 118

7   Big Data Discretization 121
    7.1   Introduction 121
    7.2   Parallel and Distributed Discretization 122
    7.3   DMDLP: Distributed Minimum Description Length Principle
        Discretizer... 123
        7.3.1   Main Discretization Procedure 125
        7.3.2   Boundary Points Selection 126
        7.3.3   DMDLP Evaluation 127
    7.4   DChi2: Distributed Chi2 Discretizer.................................. 130
    7.5   Distributed Discretization on MapReduce Associative
        Classification... 131
    7.6   A Distributed Evolutionary Multivariate Discretizer for Big
        Data.. 132
        7.6.1   Main Discretization Procedure 134
        7.6.2   Computing the Boundary Points............................. 136
        7.6.3   Distributed Cut Points Selection............................ 138
    7.7   Discretization in Big Data Streaming................................ 139
        7.7.1   IDA: Incremental Discretization Algorithm 140
        7.7.2   PiD: Partition Incremental Discretization Algorithm........ 141
        7.7.3   LOFD: Local Online Fusion Discretizer 142
    7.8   Summary and Conclusions 144
    References.. 144

8   Imbalanced Data Preprocessing for Big Data 147
    8.1   Introduction 147
    8.2   Data Sampling in Big Data.. 148
        8.2.1   ROS-BD: Random Oversampling for Big Data.............. 149
        8.2.2   RUS-BD: Random Undersampling for Big Data 150
    8.3   SMOTE-BD: SMOTE for Big Data 152
    8.4   Imbalanced Big Data Competition.................................... 154
    8.5   Other Studies on Imbalance Data Preprocessing for Big Data 155
        8.5.1   Evolutionary Undersampling 155
        8.5.2   KNN Based Data Cleaning 156

8.5.3   NRSBoundary-SMOTE 156
8.5.4   Ensemble ELM with Resampling 157
8.5.5   Imbalance Treatment for Multiclass Problems.............. 157
8.5.6   SMOTE for GPU 158
8.6   Summary and Conclusions 158
References... 158

9   Big Data Software 161
9.1   Introduction 161
9.2   MLlib: A Spark Machine Learning Library 162
9.2.1   Pipelines 164
9.2.2   Feature Extractors 164
9.2.3   Feature Transformers... 164
9.2.4   Feature Selectors 166
9.3   BigDaPSpark 166
9.3.1   Feature Selection 166
9.3.2   Data Reduction 167
9.3.3   Noise Filtering... 168
9.3.4   Missing Values Imputation................................... 170
9.3.5   Discretization 171
9.3.6   Imbalanced Learning 172
9.3.7   Random Discretization and PCA Classifier 174
9.4   FlinkML... 174
9.4.1   Data Preprocessing 175
9.4.2   Recommendation 176
9.4.3   Outlier Selection... 176
9.4.4   Utilities 176
9.5   BigDaPFlink 176
9.5.1   Feature Selection 177
9.5.2   Discretization 177
9.6   Summary and Conclusions 178
References... 179

10   Final Thoughts: From Big Data to Smart Data 183
References... 185

Acronyms

BSP	Bulk Synchronous Parallel
DAG	Directed Acyclic Graph
DM	Data Mining
FS	Feature Selection
HDFS	Hadoop Distributed File System
HPC	High-Performance Computing
IG	Instance Generation
IS	Instance Selection
KNN	K-Nearest Neighbors
ML	Machine Learning
MPI	Message Passing Interface
MV	Missing Values
PCA	Principal Components Analysis
PG	Prototype Generation
PR	Prototype Reduction
PS	Prototype Selection
RDD	Resilient Distributed Dataset
SVM	Support Vector Machine
UCI	UC Irvine Machine Learning Repository
YARN	Yet Another Resource Negotiator

Chapter 1
Introduction

1.1 Big Data

We are immersed in the Information Age where vast amounts of data are available. Petabytes of data are generated and stored everyday, resulting in a humongous *volume* of information; this information arrives at high *velocity* and its processing requires real-time processing; this information can be found in many formats, like structured, semi-structured, or unstructured data, implying *variety*; it also needs to be cleaned in order to maintain *veracity*; finally, this information must provide *value* to the organization. These five concepts are one of the most extended definitions of Big Data [1], as shown in Fig. 1.1. While the volume, velocity, and variety aspects refer to the data generation process and how to capture and store the data, veracity and value aspects deal with the quality and the usefulness of the data. These two last aspects become crucial in any Big Data process, where the extraction of useful and valuable knowledge is strongly influenced by the quality of the used data.

It is predicted that by 2020, the digital universe will be ten times as big as it was in 2013, totaling an astonishing 44 zettabytes. This explosion of data is due to three main reasons [11]: (1) thousands of applications such as sensors, social media services, and other devices that collect information continuously; (2) the improvement in storage capacity, technology, and price that has made it preferably to increase the storage rather than deleting information; (3) the improvement in machine learning (ML) approaches in the last years, enabling the acquisition of higher degree of knowledge from data.

Corporations are aware of these developments. Gaining critical business insights by querying and analyzing such massive amounts of data is becoming a necessity. This issue is known as business intelligence [39], which refers to decision support systems that combine data gathering, data storage, and knowledge management with analysis to provide input to the decision process. Regarding the former issues, a new

© Springer Nature Switzerland AG 2020
J. Luengo et al., *Big Data Preprocessing*,
https://doi.org/10.1007/978-3-030-39105-8_1

Fig. 1.1 Big Data V's

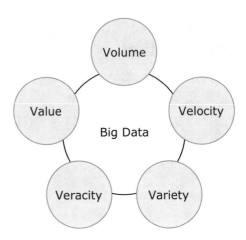

concept appears as a more general field, integrating data warehousing, data mining (DM), and data visualization for business analytics. This topic is known as data science.

The premise of Big Data is that having a world rich in data may enable machine learning and DM techniques [2] to obtain more accurate models than ever before, but classical methods fail to handle the new data space requirements. In Big Data, the usage of traditional data preprocessing techniques [17, 18] to enhance the data is even more time-consuming and resource demanding, being unfeasible in most cases. The lack of efficient and affordable data preprocessing techniques implies that the problems in data will affect the models extracted.

Even years after the boom of Big Data, there is still a misleading definition for the concept itself [16]. We must stress that the topic of Big Data is strongly linked with the scalability issue [22]. Those models developed in this context must be able to adapt dynamically the data growth, as well as being fault tolerant to be reliable in case of time-consuming operations. In order to fulfill these requirements, a change in the traditional technology and framework for carrying out the learning process is mandatory.

1.2 Big Data Analytics

Big Data Analytics is nowadays one of the most significant and profitable areas of development in data science [7, 31]. One of the main reasons of its success is related with the Internet of Things, the Web 2.0 and the social networks, and all the myriad of data from different sources that can be collected and processed [3, 6]. These technologies generate data at an exponential rate thanks to the affordability and great development of storage and network resources.

However, the real benefit of Big Data is not on the data itself, but in the ability to uncover (unexpected) patterns and glean knowledge from it with appropriate data science techniques [37]. In this sense, corporations that are able to extract valuable knowledge from large volumes of data in a reasonable time may obtain significant advantages over their competitors [8, 21]. Researchers from academia are also aware of the interest in developing robust and accurate models for DM in Big Data applications [43]. There is a clear growing rate in the number of research studies [34], and the trend is not expected to change in the short future.

The first framework that enabled the processing of large-scale datasets was MapReduce [9, 10, 33]. This revolutionary tool was intended to process and generate huge datasets in an automatic and distributed way. It is basically an execution environment which lays over a distributed file system [36]. By implementing two primitives, Map and Reduce, the user is able to use a scalable and distributed tool without worrying about technical nuances, such as failure recovery, data partitioning, or job communication.

- The Map function is devoted to divide the computation into different subparts, each one related to a partial set of the data.
- The Reduce function needs to fuse the local outputs into a single final model.

Whereas the procedure to be included in the Map task is, most times, straight-forward to determine, the hitch comes when deciding how to carry out the models fusion within the Reduce task. At this point, the design depends on many factors, namely whether the sub-models are different and independent among them, or they have a nexus for being able to join them directly.

There are different alternatives on process fusion for Big Data Analytics models under the MapReduce framework:

- Direct fusion of models: approximate methods. We refer to those that carry out a direct fusion of partial models via an ensemble system.
- Exact fusion for scalable models: distributed data and models partition. In this case, they are those designs that carry out a global distribution of both data and sub-models (the prior types mentioned just considered data division), and iteratively build the final system.

Apache Hadoop [41] emerged as the most popular open-source implementation of MapReduce, maintaining the aforementioned features. In spite of its great popularity, MapReduce (and Hadoop) is not designed to scale well when dealing with iterative and online processes, typical in ML and stream analytics [26].

Apache Spark [25] was designed as an alternative to Hadoop, capable of performing faster distributed computing by using in-memory primitives. Thanks to its ability of loading data into memory and re-using it repeatedly, this tool overcomes the problem of iterative and online processing presented by MapReduce. Additionally, Spark is a general-purpose framework that thanks to its generality allows to implement several distributed programming models on top of it (like Pregel or HaLoop) [44]. Spark is built on top of a new abstraction model called

Resilient Distributed Datasets (RDD). This versatile model allows controlling the persistence and managing the partitioning of data, among other features.

Some competitors to Apache Spark have emerged lastly, especially from the streaming side. Apache Storm [5] is an open-source distributed real-time processing platform, which is capable of processing millions of tuples per second and node in a fault-tolerant way. Apache Flink [4] is a recent top-level Apache project designed for distributed stream and batch data processing. Both alternatives try to fill the "online" gap left by Spark, which employs a mini-batch streaming processing instead of a pure streaming approach.

The performance and quality of the knowledge extracted by a DM method in any framework does not only depend on the design and performance of the method but is also very dependent on the quality and suitability of such data. Unfortunately, negative factors as noise, missing values (MV), inconsistent and superfluous data, and huge sizes in examples and features highly influence the data used to learn and extract knowledge. It is well-known that low quality data will lead to low quality knowledge [32]. Thus data preprocessing [17] is a major and essential stage whose main goal is to obtain final datasets which can be considered correct and useful for further DM algorithms.

Big Data also suffer of the aforementioned negative factors. Big Data preprocessing constitutes a challenging task, as the previous existent approaches cannot be directly applied as the size of the datasets or data streams make them unfeasible. In this book we gather the most recent proposals in data preprocessing for Big Data, providing a snapshot of the current state-of-the-art. Besides, we discuss the main challenges on developments in data preprocessing for Big Data frameworks, as well as technologies and new learning paradigms where they could be successfully applied.

1.3   Big Data Preprocessing

The set of techniques used prior to the application of a DM method is named as data preprocessing for DM [17] and it is known to be one of the most meaningful issues within the famous knowledge discovery from data process [45], as shown in Fig. 1.2. Since data will likely be imperfect, containing inconsistencies and redundancies, is not directly applicable for starting a DM process. We must also mention the fast growing of data generation rates and their size in business, industrial, academic, and science applications. The bigger amounts of data collected require more sophisticated mechanisms to analyze it. Data preprocessing is able to adapt the data to the requirements posed by each DM algorithm, enabling to process data that would be unfeasible otherwise.

Albeit data preprocessing is a powerful tool that can enable the user to treat and process complex data, it may consume large amounts of processing time [32]. It includes a wide range of disciplines, as data preparation and data reduction techniques. The former includes data transformation, integration, cleaning, and

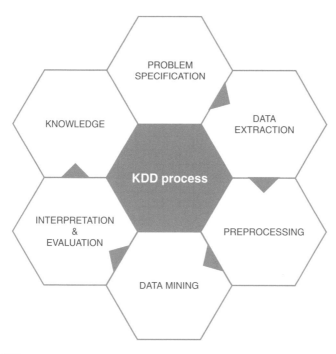

Fig. 1.2 KDD process

normalization, while the latter aims to reduce the complexity of the data by dimensionality reduction, instance reduction, or by discretization, as can be seen in Fig. 1.3. After the application of a successful data preprocessing stage, the final dataset obtained can be regarded as a reliable and suitable source for any DM algorithm applied afterwards.

Data preprocessing is not only limited to classical DM tasks, as classification or regression. More and more researchers in novel DM fields are paying increasingly attention to data preprocessing as a tool to improve their models. This wider adoption of data preprocessing techniques is resulting in adaptations of known models for related frameworks, or completely novel proposals.

Data preprocessing clearly resembles the concept of Smart Data as one of the most important stages of a DM process [23]. Its goal is to clean and correct input data, so that, a ML process may be later applied faster and with a greater accuracy. With this definition, data preprocessing techniques should enable DM algorithms to cope with Big Data problems more easily. Unfortunately, these methods are also heavily affected by the increase in size and complexity of datasets and they may be unable to provide a preprocessed/smart dataset in a timely manner, and therefore, need to be redesigned with Big Data technologies.

In the following we will present the main fields of data preprocessing, grouping them by their types and showing the current open challenges relative to each one. First, data reduction approaches will be presented, including dimension reduction,

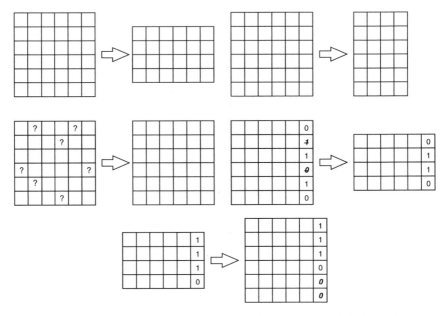

Fig. 1.3 Data preprocessing tasks, including data and data reduction, MV imputation, noise filtering, and imbalanced learning

instance reduction, and discretization. Next, we will tackle the preprocessing techniques to deal with imperfect data, where MV and noise data are included. The last section will be devoted to resampling for imbalanced problems.

1.3.1 Data Reduction

Data reduction techniques [32] emerged as preprocessing algorithms that aim to simplify and clean the raw data, enabling data mining algorithms to be applied not only in a faster way, but also in a more accurate way by removing noisy and redundant data. Data reduction methods can be divided into three different groups depending on what they reduce: dimensionality reduction, instance reduction, and discretization. From the perspective of the attributes space, the most well-known data reduction processes are feature selection (FS) and feature extraction. Regarding the reduction of the number of examples, instance reduction techniques can be divided in instance selection (IS) and instance generation methods. Finally, if the reduction affects to the values of the data points, we refer to discretization techniques.

Dimensionality Reduction

When datasets become large in the number of predictor variables or the number of instances, DM algorithms face the *curse of dimensionality* problem. It is a serious problem as it will impede the operation of most DM algorithms as the computational cost rise. This subsection will underline the most influential dimensionality reduction algorithms according to the division established into FS and feature extraction methods.

Feature Selection

FS is "the process of identifying and removing as much irrelevant and redundant information as possible" [20]. The goal is to obtain a subset of features from the original problem that still appropriately describe it. FS can remove irrelevant and redundant features which may induce accidental correlations in learning algorithms, diminishing their generalization abilities. The use of FS is also known to decrease the risk of overfitting in the algorithms used later. FS will also reduce the search space determined by the features, thus making the learning process faster and also less memory consuming.

The use of FS can also help in task not directly related to the DM algorithm applied to the data. FS can be used in the data collection stage, saving cost in time, sampling, sensing, and personnel used to gather the data. Models and visualizations made from data with fewer features will be easier to understand and to interpret.

Feature Extraction

FS is not the only way to cope with the curse of dimensionality by reducing the number of dimensions. Instead of selecting the most promising features, feature extraction techniques generate a whole new set of features by combining the original ones. Such a combination can be made obeying different criteria. The first approaches were based on linear methods, as factor analysis and Principal Components Analysis (PCA) [24].

More recent techniques try to exploit nonlinear relations among the variables. They focus on transforming the original set of variables into a smaller number of projections, sometimes taking into account the geometrical properties of clusters of instances or patches of the underlying manifolds.

Instance Reduction

A popular approach to minimize the impact of very large datasets in DM algorithms is the use of instance reduction techniques. They reduce the size of the dataset without decreasing the quality of the knowledge that can be extracted from it.

Instance reduction is a complementary task regarding FS. It reduces the quantity of data by removing instances or by generating new ones. In the following we describe the most important instance reduction and generation algorithms.

Instance Selection

Nowadays, IS is perceived as necessary [28]. The main problem in IS is to identify suitable examples from a very large amount of instances and then prepare them as input for a DM algorithm. Thus, IS is comprised by a series of techniques that must be able to choose a subset of data that can replace the original dataset and also being able to fulfill the goal of a DM application. It must be distinguished between IS, which implies a smart operation of instance categorization, from data sampling, which constitutes a more randomized approach.

A successful application of IS will produce a minimum data subset that it is independent from the DM algorithm used afterwards, without losing performance. Other added benefits of IS is to remove noisy and redundant instances (*cleaning*), to allow DM algorithms to operate with large datasets (*enabling*), and to focus on the important part of the data (*focusing*).

Instance Generation

IS methods concern the identification of an optimal subset of representative objects from the original training data by discarding noisy and redundant examples. Instance generation (IG) methods, by contrast, besides selecting data, can generate and replace the original data with new artificial data. This process allows it to fill regions in the domain of the problem, which have no representative examples in original data, or to condensate large amounts of instances in less examples. IG methods are often called prototype generation (PG) methods, as the artificial examples created tend to act as a representative of a region or a subset of the original instances.

The new prototypes may be generated following diverse criteria. The simplest approach is to relabel some examples, for example, those that are suspicious of belonging to a wrong class label. Some PG methods create centroids by merging similar examples, or by first merging the feature space in several regions and then creating a set of prototype for each one. Others adjust the position of the prototypes through the space, by adding or subtracting values to the prototype's features.

Discretization

DM algorithms require to know the domain and type of the data that will be used as input. The type of such data may vary, from categorical where no order among the values can be established, to numerical data where the order among the values there exist. Decision trees, for instance, make split based on information or separability measures that require categorical values in most cases. If continuous data is present, the discretization of the numerical features is mandatory, either prior to the tree induction or during its building process.

Discretization is gaining more and more consideration in the scientific community [27] and it is one of the most used data preprocessing techniques. It transforms quantitative data into qualitative data by dividing the numerical features into a limited number of non-overlapped intervals. Using the boundaries generated, each numerical value is mapped to each interval, thus becoming discrete. Any DM algorithm that needs nominal data can benefit from discretization methods, since many real-world applications usually produce real valued outputs. For example, three of the ten methods considered as the top ten in DM [42] need an external or embedded discretization of data: C4.5, Apriori, and Naïve Bayes. In these cases, discretization is a crucial previous stage.

Discretization also produces added benefits. The first is data simplification and reduction, helping to produce a faster and more accurate learning. The second is readability, as discrete attributes are usually easier to understand, use, and explain [27]. Nevertheless these benefits come at price: any discretization process is expected to generate a loss of information. Minimizing this information loss is the main goal pursued by the discretizer, but an optimal discretization is a NP-complete process.

1.3.2   Imperfect Data

Most techniques in DM rely on a dataset that is supposedly complete or noise-free. However, real-world data is far from being clean or complete. In data preprocessing it is common to employ techniques to either removing the noisy data or to impute (fill in) the missing data [19]. The following two subsections are devoted to MV imputation and noise filtering.

Missing Values Imputation

One big assumption made by DM techniques is that the dataset is complete. The presence of MV is, however, very common in the acquisition processes. A missing value is a datum that has not been stored or gathered due to a faulty sampling process, cost restrictions or limitations in the acquisition process. MV cannot be avoided in data analysis, and they tend to create severe difficulties for practitioners.

MV treatment is difficult. Inappropriately handling the MV will easily lead to poor knowledge extracted and also wrong conclusions [38]. MV have been reported to cause loss of efficiency in the knowledge extraction process, strong biases if the missingness introduction mechanism is mishandled and severe complications in data handling.

Many approaches are available to tackle the problematic imposed by the MV in data preprocessing. The first option is usually to discard those instances that may contain a MV. However, this approach is rarely beneficial, as eliminating instances may produce a bias in the learning process, and important information can be discarded. The seminal works on data imputation come from statistics. They model the probability functions of the data and take into account the mechanisms that induce missingness. By using maximum likelihood procedures, they sample the approximate probabilistic models to fill the MV. Since the true probability model for a particular datasets is usually unknown, the usage of ML techniques has become very popular nowadays as they can be applied without providing any prior information

Noise Treatment

DM algorithms tend to assume that any dataset is a sample of an underlying distribution with no disturbances. As we have seen in the previous section, data gathering is rarely perfect, and corruptions often appear. Since the quality of the results obtained by a DM technique is dependent on the quality of the data, tackling the problem of noise data is mandatory [12]. In supervised problems, noise can affect the input features, the output values, or both. When noise is present in the input attributes, it is usually referred as *attribute noise*. The worst case is when the noise affects the output attribute, as this means that the bias introduced will be greater. As this kind of noise has been deeply studied in classification, it is usually known as *class noise*.

In order to treat noise in DM, two main approaches are commonly used in the data preprocessing literature. The first one is to correct the noise by using *data polishing methods*, especially if it affects the labeling of an instance. Even partial noise correction is claimed to be beneficial, but it is a difficult task and usually limited to small amounts of noise. The second is to use *noise filters*, which identify and remove the noisy instances in the training data and do not require the DM technique to be modified.

1.3.3   Imbalanced Datasets

In many supervised learning applications, there is a significant difference between the prior probabilities of different classes, i.e., between the probabilities with which an example belongs to the different classes of the classification problem. This situation is known as the class imbalance problem [29]. The hitch with imbalanced datasets is that standard classification learning algorithms are often biased towards the majority class (known as the "negative" class) and therefore there is a higher misclassification rate for the minority class instances (called the "positive" examples).

While algorithmic modifications are available for imbalanced problems, our interest lies in preprocessing techniques to alleviate the bias produced by standard DM algorithms. These preprocessing techniques proceed by resampling the data to balance the class distribution. The main advantage is that they are independent of the DM algorithm applied afterwards.

Two main groups can be distinguished within resampling. The first one is *undersampling methods*, which create a subset of the original dataset by eliminating (majority) instances. The second one is *oversampling methods*, which create a superset of the original dataset by replicating some instances or creating new instances from existing ones.

Non-heuristic techniques as random oversampling or random undersampling were initially proposed, but they tend to discard information or induce overfitting. Among the more sophisticated, heuristic approaches, "Synthetic Minority Oversampling TEchnique" (SMOTE) [18] has become one of the most renowned approaches in this area. It interpolates several minority class examples that lie together. Since SMOTE can still induce overfitting in the learner, its combination with a plethora of sampling methods can be found in the specialized literature with excellent results. Undersampling has the advantage of producing reduced datasets, and thus interesting approaches based on neighborhood methods, clustering and even evolutionary algorithms have been successfully applied to generate quality balanced training sets by discarding majority class examples.

1.4   Big Data Streaming

With the advent of Big Data comes not only an increase in the volume of data, but also the notion of its velocity. In many emerging real-world problems we cannot assume that we will deal with a static set of instances. Instead, they may arrive continuously, leading to a potentially unbounded and ever-growing dataset. It will expand itself over time and new instances will arrive continuously in batches or one by one. Such problems are known as data streams [14] and pose many new challenges to DM methods. One must be able to constantly update the learning algorithm with new data, to work within time-constraints connected with the speed of arrival of instances, and to deal with memory limitations [35]. Additionally, data streams may be non-stationary, leading to occurrences of the phenomenon called concept drift, where the statistical characteristics of the incoming data may change over the time. Thus, learning algorithms should take this into consideration and have adaptation skills that allow for online learning from new instances, but also for quick changes of underlying decision mechanisms [15].

Despite the importance of data preprocessing, not many proposals in this domain may be found in the literature for online learning from data streams [46]. Most of methods are just incremental algorithms, originally designed to manage finite datasets. Direct adaptation of static data preprocessing techniques is not straightforward since most of techniques assume the whole training set is available from the beginning and properties of data do not change over time:

- Most of static IS methods require multiple passes over data, at the same time being mainly based on time-consuming neighbor searches that makes them useless for handling high-speed data streams [17].
- On the contrary, FS techniques are easily adaptable to online scenarios. Yet, they suffer from other problems such as concept evolution or dynamic and drifting feature space [30].
- Online supervised discretization methods also remain fairly unexplored. Most of standard solutions require several iterations of sharp adjustments before getting a fully operating solution [40].

Therefore, further development of data preprocessing techniques for data stream environments is thus a major concern for practitioners and scientists in DM areas.

A data stream is a potentially unbounded and ordered sequence of instances that arrive over time [13]. Therefore, it imposes specific constraints on the learning system that cannot be fulfilled by canonical algorithms from this domain. Here we list the main differences between static and streaming scenarios:

- Instances are not given beforehand, but become available sequentially (one by one) or in the form of data chunks (block by block) as the stream progresses.
- Instances may arrive rapidly and with various time intervals between each other.
- Streams are of potentially infinite size, thus it is impossible to store all of incoming data in the memory.
- Each instance may be only accessed a limited number of times (in specific cases only once) and then discarded to limit the memory and storage space usage.
- Instances must be processed within a limited amount of time to offer real-time responsiveness and avoid data queuing.
- Access to true class labels is limited due to high cost of label query for each incoming instance.
- Access to the true labels may be delayed as well, in many cases they are available after a long period, i.e., for credit approval could be 23 years.
- Statistical characteristics of instances arriving from the stream may be subject to changes over time.

Let us assume that our stream consists of a set of states $S = S_1, S_2, \ldots, S_n$, where S_i is generated by a distribution D_i. By a stationary data stream we will consider a sequence of instances characterized by a transition $S_j \rightarrow S_{j+1}$, where $D_j = D_{j+1}$. However, in most modern real-life problems the nature of data may evolve over time due to various conditions. This phenomenon is known as concept drift [15] and may be defined as changes in distributions and definitions of learned concepts over time. Presence of drift can affect the underlying properties of classes that the learning system aims to discover, thus reducing the relevance of used classifier as the change progresses. At some point the deterioration of the quality of used model may be too significant to further consider it as a meaningful component. Therefore, methods for handling drifts in data streams are of crucial importance to this area of research.

References

1. Agrawal, D., Das, S., & Abbadi, A. E. (2011). Big data and cloud computing: Current state and future opportunities. In *Proceedings of the 14th International Conference on Extending Database Technology* (pp. 530–533). New York: ACM.
2. Aha, D. W., Kibler, D., & Albert, M. K. (1999). Instance-based learning algorithms. *Machine Learning, 6*(1), 37–66.
3. Al-Fuqaha, A., Guizani, M., Mohammadi, M., Aledhari, M., & Ayyash, M. (2015). Internet of things: A survey on enabling technologies, protocols, and applications. *IEEE Communications Surveys & Tutorials, 17*(4), 2347–2376.
4. Apache Flink. (2019). http://flink.apache.org/
5. Apache Storm. (2019). https://storm.apache.org/
6. Bello-Orgaz, G., Jung, J. J., & Camacho, D. (2016). Social big data: Recent achievements and new challenges. *Information Fusion, 28*, 45–59.
7. Chen, H., Chiang, R. H. L., Storey, V. C. (2012). Business intelligence and analytics: From big data to big impact. *MIS Quarterly: Management Information Systems, 36*(4), 1165–1188.
8. Choi, T.-M., Chan, H. K., & Yue, X. (2017). Recent development in big data analytics for business operations and risk management. *IEEE Transactions on Cybernetics, 47*(1), 81–92.
9. Dean, J., & Ghemawat, S. (2008). MapReduce: Simplified data processing on large clusters. *Communications of the ACM, 51*(1), 107–113.
10. Dean, J., & Ghemawat, S. (2010). MapReduce: A flexible data processing tool. *Communications of the ACM, 53*(1), 72–77.
11. Fernández, A., del Río, S., López, V., Bawakid, A., del Jesús, M. J., Benítez, J. M. et al. (2014). Big data with cloud computing: An insight on the computing environment, MapReduce, and programming frameworks. *Wiley Interdisciplinary Reviews: Data Mining and Knowledge Discovery, 4*(5), 380–409.
12. Frénay, B., & Verleysen, M.: Classification in the presence of label noise: A survey. *IEEE Transactions on Neural Networks and Learning Systems, 25*(5), 845–869.
13. Gaber, M. M. (2012). Advances in data stream mining. *Wiley Interdisciplinary Reviews: Data Mining and Knowledge Discovery, 2*(1), 79–85.
14. Gama, J. (2010). *Knowledge discovery from data streams*. London: Chapman and Hall/CRC.
15. Gama, J., Žliobaitė, I., Bifet, A., Pechenizkiy, M., & Bouchachia, A. (2014). A survey on concept drift adaptation. *ACM computing Surveys, 46*(4), 44.
16. Gandomi, A., & Haider, M. (2015). Beyond the hype: Big data concepts, methods, and analytics. *International Journal of Information Management, 35*(2), 137–144.
17. García, S., Luengo, J., & Herrera, F. (2015). *Data Preprocessing in Data Mining*. Berlin: Springer.
18. García, S., Ramírez-Gallego, S., Luengo, J., Benítez, J. M., & Herrera, F. (2016). Big data preprocessing: Methods and prospects. *Big Data Analytics, 1*, 9.
19. García-Gil, D., Luengo, J., García, S., & Herrera, F. (2019). Enabling smart data: Noise filtering in big data classification. *Information Sciences, 479*, 135–152.
20. Hall, M. A. (1999). *Correlation-based feature selection for machine learning*. Hamilton: Department of Computer Science, Waikato University.
21. Härdle, W., Horng-Shing Lu, H., & Shen, X. (2018). *Handbook of big data analytics*. Berlin: Springer.
22. Hu, H., Wen, Y., Chua, T.-S., & Li, X. (2014). Toward scalable systems for big data analytics: A technology tutorial. *IEEE Access, 2*, 652–687.
23. Iafrate, F. (2014). A journey from big data to smart data. *Advances in Intelligent Systems and Computing, 261*, 25–33.
24. Jolliffe, I. (2011). *Principal Component Analysis*. Berlin: Springer.
25. Karau, H., Konwinski, A., Wendell, P., & Zaharia, M. (2015). *Learning spark: Lightning-fast big data analytics* (1st ed.). Sebastopol: O'Reilly Media.

26. Lin, J. (2013). MapReduce is good enough? If all you have is a hammer, throw away everything that's not a nail! *Big Data, 1*(1), 28–37.
27. Liu, H., Hussain, F., Tan, C. L., & Dash, M. (2002). Discretization: An enabling technique. *Data Mining and Knowledge Discovery, 6*(4), 393–423.
28. Liu, H., & Motoda, H. (2002). On issues of instance selection. *Data Mining and Knowledge Discovery, 6*(2), 115–130 (2002)
29. López, V., Fernández, A., García, S., Palade, V., & Herrera, F. (2013). An insight into classification with imbalanced data: Empirical results and current trends on using data intrinsic characteristics. *Information Sciences, 250*, 113–141.
30. Masud, M. M., Chen, Q., Gao, J., Khan, L., Han, J., & Thuraisingham, B. (2010). Classification and novel class detection of data streams in a dynamic feature space. In *Joint European Conference on Machine Learning and Knowledge Discovery in Databases* (pp. 337–352). Berlin: Springer.
31. Philip Chen, C. L., & Zhang, C.-Y. (2014). Data-intensive applications, challenges, techniques and technologies: A survey on big data. *Information Sciences, 275*, 314–347.
32. Pyle, D. (1999). *Data preparation for data mining*. San Francisco: Morgan Kaufmann.
33. Ramalingeswara Rao, T., Mitra, P., Bhatt, R., & Goswami, A. (2018). The big data system, components, tools, and technologies: A survey. *Knowledge and Information Systems, 60*, 1165–1245.
34. Ramírez-Gallego, S., Fernández, A., García, S., Chen, M., & Herrera, F. (2018). Big data: Tutorial and guidelines on information and process fusion for analytics algorithms with MapReduce. *Information Fusion, 42*, 51–61.
35. Ramírez-Gallego, S., Krawczyk, B., García, S., Woźniak, S., & Herrera, F. (2017). A survey on data preprocessing for data stream mining: Current status and future directions. *Neurocomputing, 239*, 39–57.
36. Shvachko, K., Kuang, H., Radia, S., & Chansler, R. (2010). The Hadoop distributed file system. In *2010 IEEE 26th Symposium on Mass Storage Systems and Technologies (MSST)* (pp. 1–10). Piscataway: IEEE.
37. Triguero, I., García-Gil, D., Maillo, J., Luengo, J., García, S., & Herrera, F. (2019). Transforming big data into smart data: An insight on the use of the k-nearest neighbors algorithm to obtain quality data. *Wiley Interdisciplinary Reviews: Data Mining and Knowledge Discovery, 9*(2), e1289.
38. Wang, H., & Wang, S. (2010). Mining incomplete survey data through classification. *Knowledge and Information Systems, 24*(2), 221–233.
39. Watson, H. J., & Wixom, B. H. (2007). The current state of business intelligence. *Computer, 40*(9), 96–99.
40. Webb, G. I. (2014). Contrary to popular belief incremental discretization can be sound, computationally efficient and extremely useful for streaming data. In *2014 IEEE International Conference on Data Mining* (pp. 1031–1036). Piscataway: IEEE.
41. White, T. (2012). *Hadoop: The Definitive Guide*. Sebastopol: O'Reilly Media.
42. Wu, X., Kumar, V., Ross Quinlan, J., Ghosh, J., Yang, Q., Motoda, H., et al. (2008). Top 10 algorithms in data mining. *Knowledge and Information Systems, 14*(1), 1–37.
43. Wu, X., Zhu, X., Wu, G.-Q., & Ding, W. (2014). Data mining with big data. *IEEE Transactions on Knowledge and Data Engineering, 26*(1), 97–107.
44. Zaharia, M., Chowdhury, M., Das, T., Dave, A., Ma, J., McCauley, M., et al. (2012). Resilient distributed datasets: A fault-tolerant abstraction for in-memory cluster computing. In *Proceedings of the 9th USENIX Conference on Networked Systems Design and Implementation* (p. 2). Berkeley: USENIX Association.
45. Zaki, M. J., & Meira, W. Jr. (2014). *Data mining and analysis: Fundamental concepts and algorithms*. New York: Cambridge University Press.
46. Zliobaite, I., & Gabrys, B. (2014). Adaptive preprocessing for streaming data. *IEEE Transactions on Knowledge and Data Engineering, 26*(2), 309–321.

Chapter 2
Big Data: Technologies and Tools

2.1 Introduction

Vast amounts of raw data are surrounding us nowadays, data that cannot be directly treated by humans or manual applications. Technologies, such as the World Wide Web, engineering and science applications and networks, business services, and many more generate data in exponential growth thanks to the development of powerful storage and connection tools. Organized knowledge and information cannot be easily obtained due to this huge data growth, neither it can be easily understood nor automatically extracted. These premises have led to the development of data science or data mining (DM) [1], a well-known discipline which is more and more present in the current world of the Information Age.

Nowadays, the current volume of data managed by our systems have surpassed the processing capacity of traditional systems [66], and this applies to DM as well. The arising of new technologies and services (like cloud computing) as well as the reduction in hardware price are leading to an ever-growing rate of information on the Internet. This phenomenon certainly represents a "Big" challenge for the data analytics community. Big Data can be thus defined as very high volume, velocity, and variety of data that require a new high-performance processing [35, 39].

Distributed computing has been widely used by data scientists before the advent of Big Data phenomenon. Many standard and time-consuming algorithms were replaced by their distributed versions with the aim of speeding up the learning process [50]. However, for most of current massive problems, a distributed approach becomes mandatory nowadays since no batch architecture is able to address such magnitudes.

Although high performance computing (HPC) tools have been used to create distributed algorithms for many years [26], HPC solutions require a high codification effort, are not naturally resilient, and any change in the hardware scheme implies a complete re-design of the algorithm. Consequently, new distributed frameworks and technologies (like Spark or Hadoop) have emerged to solve these problems [27].

© Springer Nature Switzerland AG 2020
J. Luengo et al., *Big Data Preprocessing*,
https://doi.org/10.1007/978-3-030-39105-8_2

Novel large-scale processing platforms are intended to bring closer the distributed technologies to the standard user (engineers and data scientists) by hiding the technical nuances derived from distributed environments. Complex designs are required to create and maintain these platforms, which generalizes the use of distributed computing. On the other hand, Big Data platforms also require additional algorithms that give support to the machine learning (ML) task. Standard learning algorithms must be re-designed (sometimes, entirely) if we want to learn from large-scale datasets.

Here, we classify the myriad of tools and frameworks in Big Data environment across several broad categories, according to its main role (storage, processing, high-level components), and its compatible environments (Hadoop Ecosystem, Apache Spark Framework, etc.). For instance, NoSQL databases are classified as a distributed storage component, compatible with Hadoop Ecosystem and other environments. Note that the following classification of Big Data Tools and Platforms is not an attempt to create a close taxonomy given that most of the tools are quite versatile (different uses, and work with many environments), and are constantly mutating.

In the rest of this chapter, we shall start with the discussion of common technical concepts, techniques, and paradigms which are the basement of core environments. Afterwards we shall analyze in depth the most popular frameworks in Big Data, like Hadoop or Spark, and their main components. Next we shall also discuss other novel platforms for high-speed streaming processing that are gaining importance in industry. Finally we shall make a comparison between two of the most relevant large-scale processing platforms nowadays, Spark and Flink.

2.2 Basic Concepts and Techniques

The best way of understanding the complex environment around the phenomenon of Big Data is by clearly describing the key concepts that are beneath the novel developments in this area. Starting from the standard cluster or farm of machines to cutting-edge and complex frameworks like Spark or Flink, we describe here all components that are part of Big Data solutions from many perspectives: hardware, methodologies, basic software and appliances, and so on. To better categorize these concepts, they have been divided into different sections according to the objective aimed by each one. These categories are: infrastructure, storage and access, processing, and high-level components.

2.2.1 Infrastructure

The genesis of Big Data starts with the machinery (Big Data cluster) that executes in parallel the instructions of top-level software. The cluster is logically partitioned into two types of nodes according to the main function performed: data nodes or slaves (computing) and management nodes or masters (management). Apart from its function, master and slaves can be differentiated by its computing capabilities, and its amount in the farm of nodes.

Slaves are in charge of watching partitioned data, processing and querying local data. Data locality property is here quite important, because it guarantees that our solution can scale-out by adding more resources to the cluster. Data and processing units must be as close as possible to avoid delays introduced by movements between partitions. Data nodes are normally disk-intensive, and standard in terms of computing and memory capabilities.

Masters receive and transform programs from client applications to parallel instructions that can be understand by slaves. Once client applications hit the master daemon, it eventually starts or wake several processes in the slaves that finally return an output following the opposite direction. Among the full set of responsibilities endorsed to management nodes are: failure recovery, resource management, job scheduling, monitoring, or security. To accomplish these tasks, masters require a high computing and memory power. In standard Big Data clusters, it is enough with keeping two supporting masters that watch out each other.

Both types of nodes are connected over a network connection, typically LAN (Ethernet or InfiniBand). Some configuration also allows to connect masters of different data centers across a WAN network to prevent system failures easily. In each data center, master and slaves are privately interconnected to ingest data, move data between nodes, and makes queries. There exists also another public network that acts as a facade between the client and the management service (ssh, VNC, web interface, etc.).

2.2.2 Storage and Access

Accessing Big Datum in distributed clusters is not trivial task. I/O actions must consider the availability and concurrency of every disk and processing units in the cluster. One of the most popular schemes designed to address this problem is the shared nothing architecture [61], which ensures that there are enough disk units available to supply the multiple processes running in parallel. Standard Big Data solutions use this scheme by default, though other virtualized and cloud solutions can be adopted.

No matter what model is implemented (NoSQL databases, distributed file system, SQL-based, graph-based, etc.), most of the Big Data systems rely on the idea of first bringing data into the system without spending too much effort, and then imposing an schema ("schema on read") once the data need to be accessed or processed [67]. This model is much more versatile than that implemented by relation databases since "schema on read" is loosely affected by constant updates, typically present in current dynamic systems in business and science.

Distributed file systems are by far the most popular solution for persistence in Big Data. In these systems, data are stored following the shared nothing architecture in the local disks present in the slaves (see Hadoop distributed file systems (HDFS) [34]). However, a novel approach has arisen to boost performance in storage by putting more effort in memory-based storage. Briefly, this strategy keeps in memory almost all data to be processed, and only when the processing is close to be performed data are persisted in other storage systems like HDFS, Amazon S3, or NoSQL databases.

Another concern to be addressed is the way of partitioning the data across the cluster. Given that all data cannot be saved in the same disk, and parallel processes should not access concurrently to the same disk, partitioning must be carefully tuned to fully exploit available resources in clusters. There exist multiple approaches to partition distributed data. For those schemes based on key indexing (key-value or key-tuple), two main types arise as the most popular: Range partitioning—where data are ordered and close keys are placed together—and Hash partitioning—where elements with identical keys are kept in the same partition. Apache Spark adopted both schemes by default, but there are other alternatives, such as list-based (unique values of keys are grouped in the same list), block-based (bulk storage), graph-based or round robin.

In case of failure, Big Data systems need to provide a reliable solution for failure recovery. The standard solution adopted by most of the systems is to introduce some redundancy throughout replication, and at the same time to increase its overall availability. By maintaining several copies from the same partition across different nodes and/or data centers, replication prevents from failures that might cause service interruption. Additionally, replication also increases the data locality, which in turns speeds up subsequent I/O actions. Some in-memory technologies, like Spark, propose a recovery technique based on lineage/logging [57, 68], where lost data are locally rebuilt using the logging information in case of alarm.

A myriad of input formats are currently out in the Big Data market [67]. However, the vast majority of formats can be grouped according to the stage they fit most, and some other common features. For instance, data would be typically ingested and exported/imported in delimited text format (TSV, CSV, etc.) because of its great simplicity. Formats like Parquet [49] are very suitable for column-based operations, like those present in analytics. Row-based formats, like Avro [18], become reasonable when definitions of columns are unstable, or almost every column is involved on processing.

Suitable efficient mechanisms for inter-block and intra-block indexing are mandatory in distributed computing if the requirement of fast disk accesses wants to be preserved. Two of the most popular models for fast indexing are B-Tree [23] or Bitmap Index [60]. Another possible solution to deal with indexing is to reduce the search space by specifying for each register which partitions will surely not contain it. Techniques, like Bloom Filter [21], have shown to boost performance in large-scale searching via the use of the previous schemes.

2.2.3 Distributed Processing Engines

Distributed processing frameworks in Big Data are responsible of ingesting, filtering, transforming, learning, querying, and exporting large amounts of data. Yet there exist many implementations for this concept, most of the modern distributed frameworks follow a single instruction multiple datasets [62] scheme in order to execute the same sequence of instructions simultaneously on a distributed set of data partitions. Apart from defining the main scheme of execution, these frameworks also cope with other problems, such as restarting failed processes, job scheduling, network synchronization, load-balancing, etc. In this section we focus on those models more relevant for current implementations in distributed computing.

DAG: Directed Acyclic Graph Parallel Processing

All DAG-based distributed frameworks for Big Data [25], like Spark, organize their jobs by splitting them into a smaller set of atomic tasks. In this model vertices correspond with parallel tasks, whereas edges are associated with exchange of information. As shown in Fig. 2.1, vertices can have multiple connections between inputs and outputs, which imply that the same task can be run in different data and the same data in different partitions. Data flows are physically supported by shared memory, pipes, or disks. Instructions are duplicated and sent from the master to the slave nodes for a parallel execution.

Figure 2.1 depicts a DAG execution program with 4 tasks and different partition configurations for each task. For instance, the top-most task in the graph is formed by 3 partitions. Once this task has finished, two dependent tasks (3 and 4) are started. Dependencies between partitions are not trivial as can be seen in the figure. Right-most partition in task 2 only depends on two input partitions, whereas left-most partitions has a single dependency. Finally, the input of task 4 is connected with the output of tasks 2 and 3. It is thus noteworthy to remark the differences between data (black dashed lines) and task dependencies (blue lines).

BSP: Bulk Synchronous Parallel Processing

BSP [64] systems are formed by a series of connected supersteps, implemented as directed graphs. In this scheme input data is the starting point. From here to the end, a set of supersteps are applied on partitioned data in order to obtain the final output. As mentioned before, each superstep corresponds with an independent graph associated with a subtask to be solved. Once all compounding subtasks end, bulk synchronization of all outputs is committed. At this point vertices may send messages to the next superstep, or receive some from previous steps, and also to modify its state and outgoing edges.

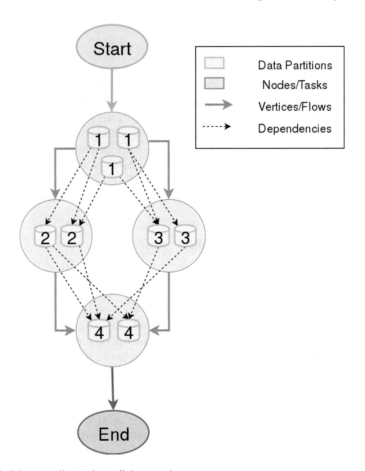

Fig. 2.1 Direct acyclic graph parallel processing

Figure 2.2 shows a toy example for BSP processing with two supersteps and one synchronization barrier. Subtasks in each superstep are depicted as rectangles with variable height (task duration), and data flows as dashed lines. Synchronization barrier acts as a time proxy between stages. Although subtask cannot start before every previous subtask has finished, however, communication between stages is allowed.

MapReduce Processing Scheme

MapReduce [24] was introduced by Google to ease the implementation of its parallel processing workflows. The main objective was to replace complex and non-intuitive programming on distributed computing (beforehand addressed by HPC platforms) by a modern transparent platform with only two functions: Map and

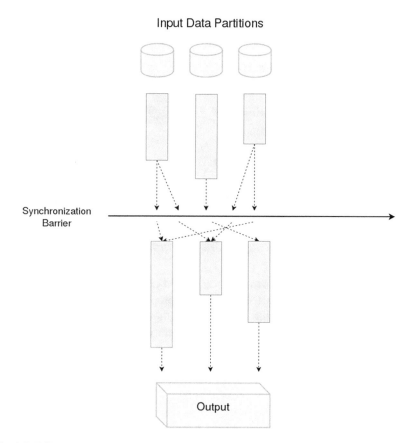

Fig. 2.2 BSP processing

Reduce. These two user-defined functions allow the users to utilize distributed resources without complaining about network, scheduling, failure recovery, etc.

Map function first reads data and transforms them into a key-value format. Transformations may apply any sequence of operations on data in a record level before sending the tuples across the network. Output keys are then grouped by key-value so that coincident keys are grouped together among with a list of values. These keys are partitioned and sent to the Reducers according to some key-based scheme previously defined (see Sect. 2.2.2). Finally, reducers typically perform some simplification on the received arrays to eventually generate tuples with a single key and value. Figure 2.3 gives an overview of the MapReduce process from a simplified perspective.

Notice that MapReduce can be deemed as a specific implementation of DAG-based processing, with only two functions as vertices. Despite these limitations, many Big Data technologies provide a wrapper implementation of MapReduce for end users (like Spark) due to its great popularity.

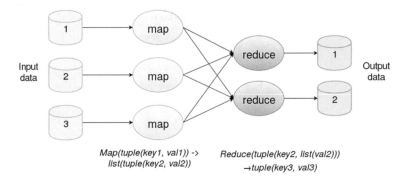

Fig. 2.3 MapReduce processing scheme

2.2.4 Other Service Management Components

On the top of processing component, there should exist another software layer that manages and checks that the system is working properly. Management services are essential for Big Data centers that can go beyond thousands of nodes, and where the failure rate is more than occasional. In this section we cover some high-level topics, like resource management or high availability.

Resource Management

Although computing and memory power are typically offered to the user as a single and global pool of resources, beneath there are a long list of single disks, processors, and memory cells to be managed. Thus, one of the major issues faced by Big Data technologies is to efficiently distribute, control, and coordinate the resources available in each machine integrated in the cluster.

In Big Data technologies, daemon processes (long-lived) or short-lived processes offer different alternatives to control resources in single nodes. Daemon processes constantly reside in the slave machines. Their job is to continuously accept and execute short tasks from the master node. In counterpart, short-lived processes are created in the slaves machines to address a specific task. Once this task has ended, the process is removed from the pool, and a new short-lived thread is added (if proceed). Sometimes these short processes can deal with many jobs if they are programmed to accept whatever is sent within a time window.

Daemon processes reduce the starting overhead since they are only created once, however, they can become idle if no request need to be addressed. In fact they can be deemed as the only valid alternative for streaming processing. On the other hand, short-lived processes act worse when many requests must be served as it increase the overall overhead time. Nevertheless they are easy to maintain and resource-efficient.

Typically, management services run in both sides: slaves and masters. The slave service informs to the master process about the slots of resources available in that node. The master service schedules the resources according to all the information collected from the subordinated services, and decides how to distribute resources.

High Availability

Two types of failure recovery must be faced by Big Data technologies to meet the high availability requirement. The first one can be provoked by the death of a management service or even the own slave node. This failure is managed by the master node by re-allocating the lost processes in another node and by restarting the dead process. The most catastrophic scenario emerges when the management process in the master dies, or its machine goes down. In this case, master node can be replaced by a secondary master node or by another strategy planned for high availability.

2.3 Hadoop Ecosystem

Undoubtedly Hadoop MapReduce can be deemed as the keystone technology in the Big Data space. After the presentation of MapReduce by Google designers [24], Hadoop MapReduce was grown by the community, and became the most used and powerful open-source implementation of MapReduce. Nowadays leading companies like Yahoo have scaled from 100-node Hadoop clusters to 42K nodes and hundreds of petabytes [36].

The main idea behind Hadoop was to create a common framework which can process large-scale data on a cluster of commodity hardware, without incurring in a high cost in developing (in contrast to HPC solutions) and execution time. Hadoop MapReduce was originally composed by two elements: the first one is the distributed storing solution called HDFS, and the second one is a data processing framework that allows to run MapReduce-like jobs. Apart from these two main goals, Hadoop implements primitives to address scalability, failure recovery, resource scheduling, etc.

But Hadoop is today more than a single technology, but a complete software stack and ecosystem formed by several components that address diverse purposes. The common factor is that all of them are built on Hadoop, and tightly depend on this technology. Some of these projects are actually Apache top-level projects [12], whereas others are continuously evolving or being created. In this section, we focus on enumerating the main components of Hadoop, as well as the cutting-edge elements recently developed for the ecosystem. All components are divided into several categories: storage, processing, high-level components. Figure 2.4 depicts a simplified scheme with the most popular components.

Fig. 2.4 Hadoop ecosystem

High-Level Components

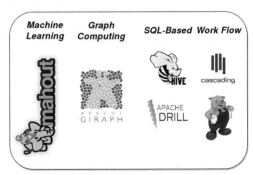

Distributed Processing Components

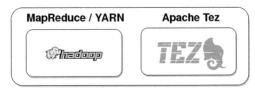

Storage Components

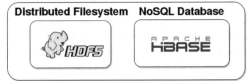

2.3.1 Distributed Storage Components

HDFS [34] is the main module of Apache Hadoop to support distributed storage. This module, which comes as a default, implements almost all concepts discussed in Sect. 2.2.2. The main abstraction here is the distributed file, which is composed by many data blocks or partitions of custom size. These partitions are distributed and allocated in the data nodes, trying to balance the disk load as much as possible. HDFS also allows replication of blocks across different nodes and racks. In HDFS, the first block is ensured to be placed in the same processing node, whereas the other two replicas are sent to different racks to prevent abrupt ends due to inter-rack issues.

HDFS was thought to work with several storage formats. It offers several APIs to read/write registers. Some relevant APIs are: InputFormat (to read customizable registers) or RecordWriter (to write record-shaped elements). Users can also develop their own storage format, and to compress data according to their requirements.

Persistence in Hadoop is mainly performed in disk. However, there are some novel advances to optimize persistence by introducing some memory usage. For instance, in Apache Hadoop version 3.0 was introduced the option of memory usage as storage medium.

Beyond distributed files, distributed storage can be addressed by non-relational databases, commonly known as NoSQL databases [56]. As the former ones were not able to adapt to high throughput problems in the Big Data scenario, NoSQL databases were created to provide simpler database designs, better scalability, and proper adaptation. Specific data structures were designed too (tuples, columns, graphs, or documents) to make NoSQL more versatile and rapid than other relational solutions for some purposes. As mentioned before, maybe the most relevant feature of NoSQL databases are its great versatility. These databases can be utilized as caching layer for high-speed data or as store for non-transactional data (web logs), among many other use cases.

In general, NoSQL databases [46] are formed by several logical layers. Logical data model layer is the main layer, which provides loosely typed data schemes for incoming data (map, column family, etc.). Data distribution layer ensures scaling-out on multiples nodes in an elastic way, maintaining at the same time the CAP theorem [33]. Persistence layer allows to save information in disk or memory or a trade-off of both. And finally, interface layer supports several non-SQL interfaces (REST, API, etc.) to access data.

MongoDB [45] is probably the most well-known open-source NoSQL database. It provides document object model for storing objects present at programming time. High-level objects are previously transformed into a JSON-based key/value tuple at runtime, and at the same time they are organized in collections (similar to tables). By doing so there is no need to define a schema, and the retrieval of a single register is much faster.

In-memory databases were designed to replace the hegemonic schema held by standard database systems until date. In-memory database systems, which focus on the utilization of memory of nodes, are intended to replace disk-based storage with fast and predictable accesses to memory. However these databases provide as an extra the feature of spill data to disk asynchronously in case of memory overflow. One of the main characteristics of in-memory solutions written in Java is the replacement of Java Heap memory by off-heap memory, not affected by GC pauses.

Some experts group in-memory solutions along with NoSQL technologies since many databases support NoSQL-based structures like columnar or document. In fact, in-memory databases meet similar use cases as NoSQL databases. They are recommended where applications require high throughput, fast analytical processing, or an intermediate temporary storage previous to the final storage. Among the long list of open-source options for memory-based storage, we can highlight some popular solutions like Hazelcast [37] or Apache Ignite [9].

2.3.2   Distributed Processing Components

Although MapReduce [24] is the native processing solution in Apache Hadoop, today it supports multiple alternatives with different processing schemes. All these solutions have in common that use a set of data nodes to run tasks on the local data blocks, and one master node (or more) to coordinate these tasks. Modern processing frameworks can be categorized into two folds according to their relationship with the Hadoop Ecosystem: the former ones are those which are exclusively designed to work within the ecosystem. Some examples are Hadoop MapReduce and Tez. Secondly, there are other technologies that are developed to run on multiple platforms, like Apache Spark and Apache Flink which can run on Hadoop. Most of these components are dependency-free, and can be interconnected easily. However some of them have acquired a major entity, and has grown in complexity and size forming a completely isolated software stack. For instance, University of Berkeley has developed a complete ecosystem around Apache Spark that deserves a complete section due to its magnitude. In the next sections, we will highlight both pluggable and complex components separately.

Apart from Hadoop MapReduce (henceforth called Hadoop MapReduce) which has been already analyzed, Hadoop's creators designed Apache Tez [16] which transforms processing jobs into DAGs. Thanks to Tez, users can run any arbitrary complex DAG of jobs in HDFS. Tez thus solve interactive and iterative processes, like those present in Hive, Pig or Cascading. Its most relevant contribution is that Tez translate any complex job to a single MapReduce phase. Furthermore, it does not need to store intermediate files and reuses idle resources, which highly boost the overall performance.

Hadoop MapReduce evolves to a more general component, called yet another resource negotiator (YARN) [17], which provides extra management and maintenance services relied to other components in the past. YARN also acts as a facade for different types of distributed processing engines based on HDFS, such as Spark, Flink, or Storm. In short, YARN was intended as a generic purpose system that separates the responsibilities of resource management (performed by YARN) and running management (performed by the given application).

Among the full set of advantages claimed by YARN, we can highlight its capacity to run several application on the same cluster without the necessity of moving data. In fact, YARN allows reusing resources across alike applications in parallel, which improves the overall usage of resources. In case of nonconformity, YARN allows users to write its own master to give full support to their requirements.

2.3.3   High-Level Components

Built on the processing layer there exist a wide range of high-level applications that gives full support to several user tasks, such as SQL-based queries, analytical processing, or ML. High-level applications translate user logic programs to a lower level so that the processing engine can understand and process them, and lastly provide trustworthy insights from data. All high-level components shown in this section are not only focused on Hadoop, namely they can work with other engines (e.g.: some algorithms in Mahout can run on Spark, or Spark can run on YARN). The opposite is also true. One distributed processing engine can give support to multiple components, which is asserted by the complete ecosystem built around Hadoop MapReduce.

In this section we outline the most popular open-source high-level applications by dividing them into several categories: SQL-based, workflow, graph processing, and ML components.

SQL-Based Components

SQL-based components are those that support ANSI SQL instructions to query and write data. In spite of relational relationships are far from appropriate in many large-scale databases and problems, they are so popular today that Big Data technologies must implement at least some basic functionalities from SQL environment. In this section, we present some open-source solutions for SQL-related problems.

Apache Hive [8] is the native data warehousing solution for Hadoop. It addresses ELT use cases (extract, load, and transform), as well as reporting through SQL interfaces. New developments in Hive add the capability of serving queries in OLAP spaces with sub-second time (Hive LLAP). One advantage of Hive is the possibility of using different processing models to solve ETL queries. In fact Hive has been tested on Tez with better performance results, and on Spark as of Hive 1.1. Hive is able to impose structure too on several input data formats like text, ORC, Parquet (columnar), or Avro.

Apache Drill [3] is SQL-based data exploration and query platform which explore data in Hadoop without requiring a formal scheme for input data. Drill utilizes a model inspired by Google's Dremel [43], which perform queries efficiently by separating out the schema from the input data. Apache Drill consists of several Drillbit services, which can act both as drivers and as slaves. Whenever a new query arrives to the system, an idle Drillbit accepts the request, creates an optimized plan for the query, and manages the entire process as leading process. The other idle Drillbit processes (determined by the Zookeeper service) receive orders from the driver Drillbit, which organizes the plan to obtain the optimal data locality.

Workflow Components

Workflow components are part of a high-level abstraction layer which allow developers to easily write processing pipelines that describe all the stages to be performed (such as read, select, aggregate, join, etc.). Workflow programs typically have their own scripting language based on popular languages like Java or Python.

All these alternatives are implemented using DAG model which allows them to yield custom and flexible programs. Here we outline the most important workflow components built for Hadoop.

Apache Pig [11] is a popular workflow component especially designed for ELT processes in data warehousing. This provides its own scripting language, called Pig Latin, to create high-level programs that can be executed on MapReduce, Tez, or even Spark. Pig can impose structure on several data formats, and connect with Hive so that both can be run in the same pipeline.

Cascading [2] is a distributed tool to create enterprise applications for Big Data without too much knowledge about MapReduce. Cascading offers a set of instructions (sinks, sources, flows, pipes, and taps) to construct Java-based pipelines for Big Data processing. It also supports platforms like Tez, Flink, or Spark.

Graph Processing Components

Graph processing components are convenient for processing data whose best representation is that based on vertices and edges instead of typical table, or array-like structures. Examples for such data are social networks, payment transactions, disease outbreaks, etc. Graph processing components are typically used with their own APIs, however, in Spark some primitives are shared among several APIs.

Apache Giraph [7] is the iterative graph processing system built in Hadoop for scalable processing of graph-based data. Inspired by Pregel architecture, Giraph utilizes the BSP strategy to process input graphs. For example, in a social network scenario, vertices represent real users, whereas edges represent friendship relations. The input for graph processing defines graph topology, and also the initial values of vertices and edges.

Machine Learning Software

Since the magnitude of learning problems has been growing exponentially, data scientists demand rapid tools that efficiently extract knowledge from large-scale data. This problem has been solved by MapReduce and other platforms by providing scalable algorithms and miscellaneous utilities in form of ML libraries. These libraries are compatible with the main Hadoop engine, and use as input the data stored in the storage components.

Apache Mahout [10] was the main contribution from Apache Hadoop to this field. Although it can be deemed as mainly obsolete nowadays, Mahout is considered as the first attempt to fill the gap of scalable ML support for Big Data. Mahout comprises several algorithms for plenty of tasks, such as classification, clustering, pattern-mining, etc. Among a long list of golden algorithms in Mahout, we can highlight Random Forest or Naïve Bayes.

The most recent version (0.13.0) provides three new major features: novel support for Apache Spark and Flink, a vector math experimentation for R, and GPU support based on large matrix multiplications. Although Mahout was originally designed for Hadoop, some algorithms have been implemented on Spark as a consequence of the latter one's popularity. Mahout is also able to run on top of Flink, being only compatible for static processing though.

2.4 Apache Spark

Apache Spark [13] was born in 2010 with the publication of the resilient distributed datasets (RDD) structures [68], the key idea behind Spark. Although Apache Spark has a close relationship with many components from Hadoop Ecosystem, Spark provides specific support for every step in the Big Data stack, such as its own processing engine, and machine learning library. In fact Apache Spark provides several persistence options to save processed data (local, network, NoSQL databases, or cloud), not originally provided by Hadoop. Figure 2.5 provides an overview of Apache Spark stack, with high-level components at the top, and some of the supported data sources and formats at the bottom.

Its outstanding popularity and remarkable differences with Hadoop promoted Apache Spark to the top-level market of Big Data platforms. The main contribution of Spark is its great capacity to rapidly process data by using in-memory distributed operations, specially designed for online and iterative computing. Some additional optimizations have also been introduced in Spark with respect to Hadoop in the shuffling and persistence stages of MapReduce jobs.

Comparing both platforms, it is easy to prove that Spark is able to complete iterative processes $100\times$ faster than Hadoop. As an extra proof, Spark won the Daytona GraySort test [59] in 2014, beating the record set by Hadoop in sorting 100 terabytes of data.

In the following, we analyze the main engine for Spark processing as well as some of the most relevant high-level components included in Apache Spark stack.

2.4.1 Spark Processing Engine

Apache Spark [38] is a distributed computing platform which can process large volume data sets in memory with a very fast response time, thanks to its memory-intensive scheme. Spark was originally thought to tackle iterative and interactive problems, which repeatedly load and/or write partial data in each interaction. Spark is compatible with several programming languages (Scala, Java, Python, and R), data storages (local file server/NFS, Cassandra, Amazon S3, etc.), and management frameworks (Tachyon, Mesos, YARN, etc.).

The heart of Spark is formed by RDD, which transparently controls how data are distributed and transformed across the cluster. Users just need to define some high-level functions that will be applied and managed by RDD. These elements are

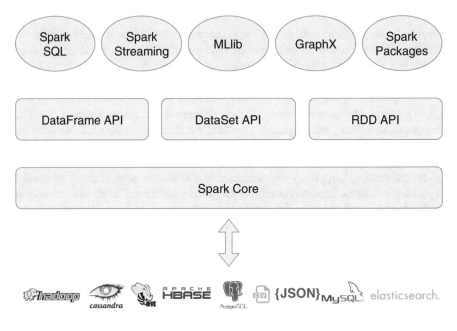

Fig. 2.5 Apache Spark framework

created whenever data are read from any source, or as a result of a transformation. RDD consists of a collection of data partitions distributed across several data nodes. A wide range of operations are provided for transforming RDD, such as filtering, grouping, set operations, among others. Furthermore RDD are also highly versatile as they allow users to customize partitioning for an optimized data placement or to preserve data in several formats and contexts.

In Spark, fault tolerance is solved by annotating operations in a structure called lineage. Spark transformations annotated in the lineage are only performed whenever trigger I/O operations appear in the log. In case of failure, Spark re-computes the affected branch in the lineage log. Although replication is normally skipped, Spark allows to spill data in local disk in case the memory capacity is not sufficient.

Spark developers provided another high-level abstraction, called DataFrames, which introduces the concept of formal schema in RDD. DataFrames are distributed and structured collections of data organized by named columns. They can be seen as a table in a relational database or a DataFrame in R, or Python (Pandas). As a plus, relational query plans built by DataFrames are optimized by the Spark's Catalyst optimizer throughout the previously defined schema. Also thanks to the scheme, Spark is able to understand data and remove costly Java serialization actions.

A compromise between structure awareness and the optimization benefits of Catalyst is achieved by the novel Dataset API. Datasets are strongly typed collections of objects connected to a relational schema. Among the benefits of datasets, we can find compile-time type safety, which means applications can be sanitized before running. Furthermore, datasets provide encoders for free to directly convert JVM objects to the binary tabular Tungsten format. These efficient in-memory formats improve memory usage and allow to directly apply operations on serialized data. Datasets are intended to be the single interface in future Spark for handling data.

Spark SQL

Spark SQL provides a single point access for efficient processing of structured data, such as Hive tables, parquet files, and JSON files. This interface helps users in accessing data using standard SQL queries. DataFrames are the main supporters of Spark SQL, though new Dataset API is also compatible with queries. Spark SQL makes easier to mix SQL queries with learning algorithms, streaming processing, or graphs.

GraphX

Likewise Spark SQL, GraphX provides a single interface for graph processing using Spark engine. This component unifies ETL, learning and exploratory analysis, and others into a single Spark API. RDD in GraphX can be viewed as both graphs and collections at the same level. Graph operators and transformations can be interchangeably performed with tiny effort. Users can also write custom iterative graph programs following the Pregel API designed by Google. GraphX includes several golden algorithms for graph processing like PageRank or Label Propagation.

Spark Streaming

Spark Streaming provides stateful common semantics to write streaming jobs for Spark engine. This API enables scalable, high throughput, fault-tolerant stream processing of unbounded data streams. Input can be ingested from many sources like Kafka, Flume, or TCP sockets, and can be processed using complex functions like map, join, and window. As part of Spark, code for processing and ingesting streaming events may be integrated with SQL queries, predictive models, etc.

Internally, Spark Streaming receives unbounded data and divides them into discrete batches, which are later processed by the Spark engine to generate a stream of results also represented as batches. Discretized Streams or DStreams are the main structure used to represent these streams in form of RDD. Each RDD is formed by data from a certain discrete interval. Any transformation applied on a DStream translates to individual operations on the underlying RDD.

2.4.2 MLlib

MLlib project [44] was born in 2012 as an extra component of Spark. It was released and open-sourced in 2013 under the Apache 2.0 license. From its inception, the number of contributions and people involved in the project has been growing steadily. Apart from official API, Spark provides a community package index [58] (Spark Packages) to assemble all open-source algorithms that work with MLlib.

MLlib is a Spark library geared towards offering distributed ML support to Spark engine. This library includes several out-of-the-box algorithms for alike tasks, such as classification, clustering, regression, recommendation, even data preprocessing. Apart from distributed implementations of standard algorithms, MLlib offers:

- Common utilities: for distributed linear algebra, statistical analysis, internal format for model export, data generators, etc.
- Algorithmic optimizations: from the long list of optimizations included, we can highlight some: decisions trees, which borrow some ideas from PLANET project [47] (parallelized learning both within trees and across them); or generalized linear models, which benefit from employing fast C++-based linear algebra for internal computations.
- Pipeline API: as the learning process in large-scale datasets is tedious and expensive, MLlib includes an internal package (*spark.ml*) that provides an uniform high-level API to create complex multi-stage pipelines that connect several and alike components (preprocessing, learning, evaluation, etc.). *spark.ml* allows model selection or hyper-parameter tuning, and different validations strategies like k-fold cross validation.
- Spark integration: MLlib is perfectly integrated with other Spark components. Spark GraphX has several graph-based implementations in MLlib, like LDA. Likewise, several algorithms for online learning are available in Spark Streaming, such as online k-Means. In any case, most of the components in the ASF stack are prepared to effortlessly cooperate with MLlib.

MLlib also includes several of the most popular and widely used algorithms for classification and regression.

Logistic Regression

Logistic regression is a linear method that predicts a categorical response using the probability. The loss function is given by the logistic loss. It can be used for both binary problems (binomial logistic regression) and multiclass problems (multinomial logistic regression).

Spark's implementation of the logistic regression algorithm uses the limited-memory BFGS (L-BFGS) algorithm [40] for optimization of the memory used.

Decision Tree

Decision trees are one of the most popular methods in ML for both classification and regression tasks. They are easy to interpret, can handle categorical features, and extend to the multiclass classification problem among other features.

A decision tree uses a tree-like graph for decision-making [54]. It starts with a single node which divides into possible outcomes. Each of those outcomes leads to additional nodes, which in turn are divided into other nodes. The end nodes are the decision of a certain branch of the tree.

Spark's implementation of the decision tree is optimized for scalability. The key optimizations are: level-wise training, for selecting the splits for all nodes at the same level of the tree, approximate quantiles, bin-wise computation, for saving computation on each iteration by precomputing the binned representations of each instance, and the avoiding of the map operation.

Random Forest

Ensembles are algorithms that combine a set of models build upon other machine learning algorithms. Random Forests are a combination of decision trees where each tree is trained independently using a random sample of the data.

Spark's Random Forest implementation builds upon the decision tree code, which distributes the learning of single trees. Many of the optimizations are based upon Google's PLANET project [48]. Random Forests are easily paralleled since each tree can be trained independently. Spark's Random Forest does exactly that, a variable number of subtrees are trained in parallel.

GBTs: Gradient-Boosted Trees

Like Random Forest, GBTs are also ensembles of decision trees. Decision trees are trained iteratively, minimizing a loss function. Each iteration, the algorithm uses the current model to predict the label of each instance in the train set. Then it compares the predicted label with the real one. The dataset is then relabeled to put more emphasis on the instances with bad predictions. The next iteration, the trained decision tree will help with the wrong predictions [29].

GBTs can be employed for both binary classification and regression, using both continuous and categorical features.

MLP: MultiLayer Perceptron

MLP classifiers are based on the feedforward artificial neural network [53]. They are composed of multiple layers fully connected to each other. Every node in the input layer represents the input data. All nodes transform the input data by a linear combination with the nodes.

SVM: Linear Support Vector Machine

SVM builds a hyperplane (or a set of hyperplanes) in a high-dimensional space. They can be used for classification, regression, or other tasks. Spark's implementation of SVM optimizes the hinge loss [52] using OWLQN optimizer.

OVA: One-vs-All

OVA is a popular strategy for multiclass classification when using a binary classifier [30]. OVA learns a model from each class, considering the instances of that class as positive samples, and the rest of the instances as negative samples.

Naïve Bayes

Naïve Bayes is a simple probabilistic classifier for multiclass problems based on the Bayes theorem [55]. It constructs rules that will be used for assigning examples to a certain class, while assuming independence between every pair of attributes.

Naïve Bayes is a very efficient algorithm as it only needs a single pass over the training data for computing the conditional probability distribution of each attribute for a certain label.

MLlib supports both Bernoulli and multinomial Naïve Bayes.

Linear Regression

Linear regression is a statistic technique employed for studying the relation between variables. There are two types of linear regressions: simple linear regression in the case of one explanatory variable and multiple linear regression in the presence of two or more explanatory variables.

Generalized Linear Regression

Contrary to the linear regression, where the response is assumed to follow a Gaussian distribution, the generalized linear regression follows an exponential distribution.

Apache Spark's implementation of generalized linear regression also provides summary statistics, including p-values, residuals, Akaike information criterion, and deviances, among others.

Survival Regression

MLlib implements the accelerated failure time (AFT) model [65], which is used to analyze survivorship data.

Different from a proportional hazards model [28] designed for the same purpose, the AFT model is easier to parallelize because each instance contributes to the objective function independently.

Isotonic Regression

Isotonic (or monotonic) regression [19] is the technique of fitting a free-form line to a sequence of observations. However, it has certain constraints: the fitted free-form line must be non-decreasing (or non-increasing) everywhere, and it must lie as close to the observations as possible.

2.4.3 Spark Packages

Spark Packages is an open-source community package index for Apache Spark [58]. It keeps track of the different libraries and packages created for improving the Apache Spark ecosystem. Currently, it contains more than 400 packages, categorized depending on the purpose of the package.

ML is one of the most prolific categories of Spark Packages, containing almost 100 contributions. These packages extend Apache Spark's MLlib with new algorithms. Here we present two elemental algorithms used for preprocessing data:

- KNN_IS: The K-nearest neighbors (KNN) algorithm experiences a series of difficulties to deal with big datasets, such as high computational cost, high storage requirements, sensitivity to noise, and inability to work with incomplete information. In Big Data environments, Maillo et al. [42] proposed a technological solution based on Apache Spark for the standard KNN algorithm to partly alleviate some of problems stated above (memory consumption and computation cost) by means of a distributed computation of nearest neighbors. In Fig. 2.6 we

SparkPackages   KNN_IS

This is an open-source Spark package about an exact k-nearest neighbors classification based on Apache Spark. We take advantage of its in-memory operations to simultaneously classify big amounts of unseen cases against a big training dataset. The map phase computes the k-nearest neighbors in different splits of the training data. Afterwards, multiple reducers process the definitive neighbors from the list obtained in the map phase. The key point of this proposal lies on the management of the test set, maintaining it in memory when it is possible. Otherwise, this is split into a minimum number of pieces, applying a MapReduce per chunk, using the caching skills of Spark to reuse the previously partitioned training set.

```
spark-shell --packages JMailloH:KNN_IS:3.0
```

https://spark-packages.org/package/JMailloH/KNN_IS

Fig. 2.6 Spark package: KNN_IS

SparkPackages   spark-knn

k-Nearest Neighbors algorithm implemented on Apache Spark. This uses a hybrid spill tree approach to achieve high accuracy and search efficiency. It scales very well both horizontally and in terms of number of observations/dimensions.

```
spark-shell --packages saurfang:spark-knn:0.2.0
```

https://spark-packages.org/package/saurfang/spark-knn

Fig. 2.7 Spark package: KNN hybrid spill tree

can find a Spark Package associated with this research in the third-party Apache Spark Repository.

- spark-knn: In [41], a hybrid spill tree is proposed to compute parallel KNN, hybridizing metric trees and spill trees to speed up the classification and maintain a good performance. This approximate approach dramatically reduces the computational costs of the KNN algorithm in a Big Data context with a high number of instances. In Fig. 2.7 we can find an open-source implementation of this hybrid spill tree available as a Spark Package.

Spark Packages also provides a way to interconnect other frameworks with Apache Spark. For example, *spark-sklearn*[1] integrates the Spark computing framework with the popular scikit-learn library [22]. *CaffeOnSpark*[2] enables deep learning in Spark clusters. Finally, *sparkling-water*[3] connects worlds of H2O [63] and Spark.

2.5 Streaming Processing Frameworks for Big Data

Streaming processing frameworks are those platforms mainly designed to process unbounded and continuous streams of data in controlled time periods. These systems typically process streams in forms of batches or windows, which are aggregated to provide partial results in each time.

The usual model implemented by these systems is the producer–consumer agent model, where each agent intercepts streaming data, processes data, and eventually sends the result to the next element. Agents are networked to implement complex logic models and relationships.

Streaming processing technologies are normally integrated with other frameworks (e.g., Hadoop, Spark, etc.), as extra services. These systems are only responsible of processing streaming events and providing results to other external services.

2.5.1 *Apache Flink*

Apache Flink [4] is a distributed processing component focused on streaming processing, which was designed to solve problems derived from micro-batch models (Spark Streaming). Flink also supports batch data processing with programming abstractions in Java and Scala, though it is treated as a special case of streaming processing. In Flink, every job is implemented as a stream computation, and every task is executed as cyclic data flow with several iterations. Flink provides two operators for iterations [5], namely standard and delta iterator. In standard iterator, Flink only works with a single partial solution, whereas delta iterator utilizes two worksets: the next entry set to process and the solution set.

Apache Flink offers a high fault tolerance mechanism to consistently recover the state of data streaming applications. This mechanism is generating consistent snapshots of the distributed data stream and operator state. In case of failure, the system can fall back to these snapshots.

---

[1]https://spark-packages.org/package/databricks/spark-sklearn.

[2]https://spark-packages.org/package/yahoo/CaffeOnSpark.

[3]https://spark-packages.org/package/h2oai/sparkling-water.

Apache Flink has four big libraries built on those main APIs:

- Gelly: is the graph processing system in Flink. It contains methods and utilities for the development of graph analytic applications.
- FlinkML: this library aims to provide a set of scalable ML algorithms and an intuitive API. Until now, FlinkML provides few alternatives for some fields in machine learning: SVM with CoCoA or multiple linear regression for supervised learning, KNN join for unsupervised learning, scalers and polynomial features for preprocessing, alternating least squares for recommendation, and other utilities for validation and outlier selection, among others. FlinkML also allows users to build complex analysis pipelines via chaining operations (like in MLlib). These pipelines are inspired by the design introduced by scikit-learn in [22]. FlinkML is described in detail in Sect. 9.4.
- Table API and SQL: is a SQL-like expression language for relational stream and batch processing that can be embedded in Flink's data APIs.
- FlinkCEP: is the complex event processing library. It allows to detect complex events patterns in streams.

FlinkML also includes some of the most popular and widely used algorithms for supervised and unsupervised learning.

Supervised Learning

FlinkML includes two algorithms for supervised learning. It also comes with an optimization framework for finding the best parameters for a certain model:

- SVM with CoCoA: it is an SVM classifier using a linear optimizer. The communication-efficient distributed dual coordinate ascent algorithm (CoCoA) [32] and the stochastic dual coordinate ascent (SDCA) algorithms are used in Flink to solve the previously defined minimization problem. CoCoA consists of several iterations of SDCA on each partition, and a final phase of aggregation of partial results. The result is a final gradient state, which is replicated across all nodes and used in further steps.
- Multiple linear regression: it uses stochastic gradient descent (SGD) [51] to approximate the gradient solutions with squared loss. In SGD a sample of data (called mini batches) is used to compute subgradients in each phase. Only the partial results from each worker are sent across the network in order to update the global gradient.
- Optimization framework: FlinkML also provides an optimization framework. This framework is focused on supervised learning. It can be used to find a model, defined by a set of parameters that minimize a function given a set of labeled examples. It supports different loss functions (squared, hinge, and logistic loss) as well as two regularization types (L_1 and L_2). It also allows the use of SGD for finding the local minimum of the function.

Unsupervised Learning

The unsupervised section of FlinkML is completed with an exact KNN join algorithm [20].

This algorithm uses a brute-force approach for computing the distance between every pair of training and test examples. For speeding up the computation of the distances, the algorithm uses a quadtree. The quadtree is very efficient with many examples, but scales poorly with the increasing number of dimensions. That is why the algorithm will automatically choose whether or not to use the quadtree.

2.5.2 Apache Flume

Apache Flume [6] was originally developed for processing application logs; however, the current version of Flume supports any type of streaming data. Flume is based on four key concepts: source, sink, channel, and interceptors. Whereas sources are intended to be the origin of streaming data, sinks act as the former destinations for processed data. Channels are arbitrary systems in charge of transferring data from sources to sinks. Lastly, interceptors alter or remove moving events.

Flume comprises multiple sources, sinks, channels, and configurations to create complex data flow pipelines for processing purposes. Flume is specially thought for simple streaming use cases (filtering or aggregations) since these features are natively implemented in Flume. For complex problems, Flume also allows custom code. Easy deploy and configuration of Flume is another valuable advantage of its use.

2.5.3 Apache Storm

Apache Storm [15] is another distributed processing framework for processing high volume of unbounded streaming events in real time. Storm is a versatile tool that offers support for several use cases (analytics, ML, unbounded computation) and data sources (HDFS, Cassandra, etc.). Storm is much more suitable for complex processing requirements than Flume thanks to its complete Trident API, full of operators.

The core abstractions in Storm are: the Stream (represents an unbounded sequence of tuples), the Spouts (read events from a source), and the Bolts (consumes input streams, process them, and emits transformed streams). All these components are connected forming a network (Topology) in order to run complex use cases of stream transformations.

Resulting topologies are submitted to storm clusters for execution. Edges in topology represent the subscriptions of Bolts to one or more nodes (Bolt or Spout). When a node emits some tuples, they are sent to every Bolt that subscribes to the stream part of the node.

2.6 Spark vs. Flink: A Thorough Comparison Between Two Outstanding Platforms

In this section, we make a thorough comparison between two of the most promising distributing processing engines for large-scale ML: Apache Spark and Apache Flink [31]. The main differences and analogies between both engines are presented in order to outline which are the best scenarios for each platform.

The first remarkable difference between both frameworks lies in the way each tool ingests streams of data. Whereas Flink is a native streaming processing framework that can work with batch data, Spark was originally designed to work with static data through its RDD. Spark uses micro-batching to deal with streams. Micro-batching divides incoming data and processes small parts one at a time. Its main advantage is that the structure chosen by Spark, called DStream, is a simple queue of RDD. This approach allows to easily interleave between streaming and batch APIs. However, micro-batching are not well prepared for systems that require very low latency. Nevertheless, Flink fits perfectly well in those systems as it natively uses streams for all type of workloads [32].

Unlike Hadoop MapReduce, Spark and Flink have support for data re-utilization and iterations. Spark keeps data in memory across iterations through an explicit caching. However, Spark plans its executions as acyclic graph plans, which implies that it needs to schedule and run the same set of instructions in each iteration. In contrast, Flink implements a thoroughly iterative processing in its engine based on cyclic data flows (one iteration, one schedule). Additionally, it offers delta iterations to leverage operations that only changes part of data.

Till the advent of Tungsten optimization project, Spark mainly used the JVM's heap memory to store objects [14]. Although it is straightforward solution, it may suffers from overflow memory problems and GC pauses. Thanks to Tungsten, these problems started to disappear. DataFrames allows Spark to manage its own memory stack and to exploit the memory hierarchy available in modern computers (L1 and L2 CPU caches). Flink's designers, however, had these facts into consideration from the early stages [5]. Flink's team thus proposed to maintain a self-controlled memory stack, with its own type extraction and serialization strategy based on binary format. The advantages derived from these tunes are: less memory errors, less GC pressure, and a better space data representation, among others.

Regarding optimization, both frameworks have mechanisms that analyze the code submitted by the user and yield the best pipeline code for a given execution graph; Spark through the DataFrames API and Flink as first citizen. For instance, in Flink a join operation can be planned as a complete shuffling of two sets, or as a broadcast of the smallest one. Spark also offers a manual optimization, which allows the user to control partitioning and memory caching.

References

1. Aggarwal, C. C. (2015). *Data mining: The textbook*. Berlin: Springer.
2. Apache Cascading. (2019). http://www.cascading.org/
3. Apache Drill. (2019). Apache Drill. https://drill.apache.org/
4. Apache Flink. (2019). http://flink.apache.org/
5. Apache Flink Project. (2015). Peeking into Apache Flink's Engine Room. https://flink.apache.org/news/2015/03/13/peeking-into-Apache-Flinks-Engine-Room.html
6. Apache Flume. (2019). https://flume.apache.org/
7. Apache Giraph. (2019). Apache Giraph. https://giraph.apache.org/
8. Apache Hive. (2019). https://hive.apache.org/
9. Apache Ignite. (2019). https://ignite.apache.org/
10. Apache Mahout. (2019). https://mahout.apache.org/
11. Apache Pig. (2019). https://pig.apache.org/
12. Apache Software Foundation. (2019). Apache project directory. https://projects.apache.org/
13. Apache Spark. (2019). Apache Spark: Lightning-fast cluster computing. http://spark.apache.org/
14. Apache Spark Project. (2015). Project Tungsten (Apache Spark). https://databricks.com/blog/2015/04/28/project-tungsten-bringing-spark-closer-to-bare-metal.html
15. Apache Storm. (2019). https://storm.apache.org/
16. Apache Tez. (2019). https://tez.apache.org/
17. Apache YARN. (2019). https://hadoop.apache.org/docs/current/hadoop-yarn/hadoop-yarn-site/YARN.html
18. Avro Project. (2019). *Avro Project*. https://avro.apache.org/
19. Barlow, R. E., & Brunk, H. D. (1972). The isotonic regression problem and its dual. *Journal of the American Statistical Association, 67*(337), 140–147.
20. Böhm, C., & Krebs, F. (2004). The k-nearest neighbour join: Turbo charging the KDD process. *Knowledge and Information Systems, 6*(6), 728–749.
21. Broder, A., & Mitzenmacher, M. (2004). Network applications of bloom filters: A survey. *Internet Mathematics, 1*(4), 485–509.
22. Buitinck, L., Louppe, G., Blondel, M., Pedregosa, F., Mueller, A., Grisel, O., et al. (2013). API design for machine learning software: Experiences from the scikit-learn project. In *ECML PKDD Workshop: Languages for Data Mining and Machine Learning* (pp. 108–122).
23. Comer, D. (1979). Ubiquitous B-tree. *ACM Computing Surveys, 11*(2), 121–137.
24. Dean, J., & Ghemawat, S. (2004). MapReduce: Simplified data processing on large clusters. In *OSDI04: Proceedings of the 6th Conference on Symposium on Operating Systems Design and Implementation*. Berkeley: USENIX Association.
25. Dennis, J. B. (1974). *First version of a data flow procedure language* (pp. 362–376). Berlin: Springer.
26. Dursi, J. (2019). *HPC is dying, and MPI is killing it*. https://www.dursi.ca/post/hpc-is-dyingand-mpi-is-killing-it.html/. Online; accessed July 2019.
27. Fernández, A., del Río, S., López, V., Bawakid, A., del Jesús, M. J., Benítez, J. M., et al. (2014). Big data with cloud computing: An insight on the computing environment, MapReduce, and programming frameworks. *Wiley Interdisciplinary Reviews: Data Mining and Knowledge Discovery, 4*(5), 380–409.
28. Fine, J. P., & Gray, R. J. (1999). A proportional hazards model for the subdistribution of a competing risk. *Journal of the American Statistical Association, 94*(446), 496–509.
29. Friedman, J. H. (2001). Greedy function approximation: A gradient boosting machine. *Annals of Statistics, 29*, 1189–1232.
30. Galar, M., Fernández, A., Barrenechea, E., Bustince, H., & Herrera, F. (2011). An overview of ensemble methods for binary classifiers in multi-class problems: Experimental study on one-vs-one and one-vs-all schemes. *Pattern Recognition, 44*(8), 1761–1776.

31. García, S., Ramírez-Gallego, S., Luengo, J., Benítez, J. M., & Herrera, F. (2019). Big data preprocessing: Methods and prospects. *Big Data Analytics, 1*(1), 9.
32. García-Gil, D., Ramírez-Gallego, S., García, S., & Herrera, F. (2017). A comparison on scalability for batch big data processing on Apache Spark and Apache Flink. *Big Data Analytics, 2*(1), 1.
33. Gilbert, S., & Lynch, N. (2012). Perspectives on the cap theorem. *Computer, 45*(2), 30–36.
34. Hadoop Distributed File System. (2019). https://hadoop.apache.org/docs/stable/hadoop-project-dist/hadoop-hdfs/HdfsUserGuide.html
35. Härdle, W., Horng-Shing Lu, H., & Shen, X. (2018). *Handbook of big data analytics*. Berlin: Springer.
36. Harris, D. (2013). The history of Hadoop: From 4 nodes to the future of data. https://gigaom.com/2013/03/04/the-history-of-hadoop-from-4-nodes-to-the-future-of-data/
37. Hazelcast. (2019). https://hazelcast.com/
38. Karau, H., Konwinski, A., Wendell, P., & Zaharia, M. (2015). *Learning spark: Lightning-fast big data analytics*. Sebastopol: O'Reilly Media.
39. Laney, D. (2001). *3D data management: Controlling data volume, velocity and variety*. http://blogs.gartner.com/doug-laney/files/2012/01/ad949-3D-Data-Management-Controlling-Data-Volume-Velocity-and-Variety.pdf. Online; accessed March 2019.
40. Liu, D.C., & Nocedal, J. (1989). On the limited memory BFGS method for large scale optimization. *Mathematical Programming, 45*(1), 503–528.
41. Liu, T., Rosenberg, C. J., & Rowley, H. A. (2009). Performing a parallel nearest-neighbor matching operation using a parallel hybrid spill tree. US Patent 7,475,071.
42. Maillo, J., Ramírez, S., Triguero, I., & Herrera, F. (2016). kNN-IS: An iterative spark-based design of the k-nearest neighbors classifier for big data. *Knowledge-Based Systems, 117*, 3–15.
43. Melnik, S., Gubarev, A., Long, J. J., Romer, G., Shivakumar, S., Tolton, M., et al. (2010). Dremel: Interactive analysis of web-scale datasets. In *Proceedings of the 36th International Conference on Very Large Data Bases* (pp. 330–339).
44. Meng, X., Bradley, J., Yavuz, B., Sparks, E., Venkataraman, S., Liu, D., et al. (2016). MLlib: Machine learning in apache spark. *Journal of Machine Learning Research, 17*(34), 1–7.
45. MongoDB. (2019). https://www.mongodb.com/
46. NoSQL Database. (2019). NoSQL database. http://nosql-database.org/
47. Panda, B., Herbach, J. S., Basu, S., & Bayardo, R. J. (2009). Planet: Massively parallel learning of tree ensembles with MapReduce. In *Proceedings of the 35th International Conference on Very Large Data Bases (VLDB-2009)*.
48. Panda, B., Herbach, J. S., Basu, S., & Bayardo, R. J. (2009). Planet: Massively parallel learning of tree ensembles with MapReduce. *Proceedings of the VLDB Endowment, 2*(2), 1426–1437.
49. Parquet Project. (2019). *Parquet Project*. https://parquet.apache.org/
50. Ramalingeswara Rao, T., Mitra, P., Bhatt, R., & Goswami, A. (2019). The big data system, components, tools, and technologies: A survey. *Knowledge and Information Systems, 60*, 1165–1245.
51. Robbins, H., & Monro, S. (1985). A stochastic approximation method. In *Herbert Robbins selected papers* (pp. 102–109). Berlin: Springer.
52. Rosasco, L., De Vito, E., Caponnetto, A., Piana, M., & Verri, A. (2004). Are loss functions all the same? *Neural Computation, 16*(5), 1063–1076.
53. Rosenblatt, F. (1961). Principles of neurodynamics. Perceptrons and the theory of brain mechanisms. Technical report, Cornell Aeronautical Lab Inc., Buffalo.
54. Ross Quinlan, J. (1986). Induction of decision trees. *Machine Learning, 1*(1), 81–106.
55. Russell, S. J., & Norvig, P. (2016). *Artificial intelligence: a modern approach*. Kuala Lumpur: Pearson Education Limited.
56. Sadalage, P. J., & Fowler, M. (2012). *NoSQL distilled: A brief guide to the emerging world of polyglot persistence*. Addison-Wesley Professional (1st ed.). Boston: Addison-Wesley.
57. Simmhan, Y. L., Plale, B., & Gannon, D. (2005). A survey of data provenance techniques. Technical Report, Indiana University.
58. Spark Packages. (2019). 3rd party spark packages. https://spark-packages.org/

59. Spark Petabyte Sort. (2014). Apache Spark the fastest open source engine for sorting a petabyte. https://databricks.com/blog/2014/10/10/spark-petabyte-sort.html
60. Spiegler, I., & Maayan, R. (1985). Storage and retrieval considerations of binary data bases. *Information Processing and Management, 21*(3), 233–254.
61. Stonebraker, M. (1986) The case for shared nothing. *Database Engineering, 9*, 4–9.
62. Sung, M. (2000). SIMD parallel processing Michael Sung 6.911: Architectures anonymous. http://www.ai.mit.edu/projects/aries/papers/writeups/darkman-writeup.pdf/. [Online; accessed July 2019].
63. The H2O.ai team. (2015). *H2O: Scalable machine learning*. http://www.h2o.ai
64. Valiant, L. G. (1990). A bridging model for parallel computation. *Communications of ACM, 33*(8), 103–111.
65. Wei, L.-J. (1992). The accelerated failure time model: A useful alternative to the Cox regression model in survival analysis. *Statistics in Medicine, 11*(14–15), 1871–1879.
66. Wu, X., Zhu, X., Wu, G.-Q., & Ding, W. Data mining with big data. *IEEE Transactions on Knowledge and Data Engineering, 26*(1), 97–107.
67. Yu, S., & Guo, S. (2016). *Big data concepts, theories, and applications*. Amsterdam: Elsevier.
68. Zaharia, M., Chowdhury, M., Das, T., Dave, A., Ma, J., McCauley, M., et al. (2012). Resilient distributed datasets: A fault-tolerant abstraction for in-memory cluster computing. In *Proceedings of the 9th USENIX Conference on Networked Systems Design and Implementation, NSDI'12* (pp. 2–2).

Chapter 3
Smart Data

3.1 Introduction

Data is the natural resource of the twenty-first century. Such resource is vastly gathered, resulting in huge amounts of data that are stored every second. These high dimensional data pools, along with their associated technologies, are often referred to as Big Data. Big Data as concept is defined around five aspects [12]: data volume, data velocity, data variety, data veracity, and data value. While the volume, variety, and velocity aspects refer to the data generation process and how to capture and store the data, veracity and value aspects deal with the quality and the usefulness of the data. These two last aspects become crucial in any Big Data [16], where the extraction of useful and valuable knowledge is strongly influenced by the quality of the used data. Nowadays, Big Data as a discipline has already gone through its implantation cycle and is nowadays transversal for any data driven enterprise. Thus, Big Data solutions and technologies represent the current way to reach competitiveness and growth in our society, where connected devices are rapidly increasing, generating data at ever growing rates.

Recently, Smart Data (focusing on veracity and value) has been introduced, aiming to filter out, or amend, the imperfections and to highlight the valuable data, which can be effectively used by companies and governments for planning, operation, monitoring, control, and intelligent decision-making. Three key attributes are needed for data to be smart, it must be accurate, actionable, and agile:

- Accurate: data must be what it says it is with enough precision to drive value. Data quality matters.
- Actionable: data must drive an immediate scalable action in a way that maximizes a business objective like media reach across platforms. Scalable action matters.
- Agile: data must be available in real time and ready to adapt to the changing business environment. Flexibility matters.

© Springer Nature Switzerland AG 2020
J. Luengo et al., *Big Data Preprocessing*,
https://doi.org/10.1007/978-3-030-39105-8_3

A challenge that becomes even trickier is the management of the quality of the data in Big Data environments. Advanced Big Data modeling and analytics are indispensable for discovering the underlying structure from retrieved data in order to acquire Smart Data. Data must be appropriately sorted, structured, analyzed to get smarter analysis and be usable in the future; make data actionable in order to address customer and business challenges putting data in the context of purpose [3]. In this book we provide several preprocessing techniques for Big Data, transforming raw, corrupted datasets into Smart Data.

In the rest of this chapter we shall discuss the state of Smart Data in the literature, differentiating the "Big" component of the data from the "Smart" part of it.

3.2   From Big Data to Smart Data

Big Data is an appealing discipline that presents an immense potential for global economic growth and promises to enhance competitiveness of high technological countries. As in any knowledge extraction process, vast amounts of data are analyzed, processed, and interpreted in order to generate profits in terms of either economic or advantages for society. Once the Big Data has been analyzed, processed, interpreted, and cleaned, it is possible to access it in a structured way. This transformation is the difference between "Big" and "Smart" Data [15], as can be seen in Fig. 3.1.

The first step in this transformation is to perform an integration process, where the semantics and domains from several large sources are unified under a common structure. The usage of ontologies to support the integration is a recent approach [4], but graph databases are also an option where the data is stored in a relational form, as in healthcare domains [20].

Even when the integration phase ends, the data is still far from being "smart": the accumulated noise in Big Data tasks creates problems, especially when the dimensionality is large [6]. As data grows, noise accumulates and algorithmic instability appears [7], particularly when a massive sample pool has been integrated from heterogeneous sources. Thus, in order to be "smart," the data still needs to be cleaned even after its integration. Unfortunately, there is a lack of proposals for noise cleaning in Big Data environments. Since data cleaning usually imposes the creation of several classifiers and data partitioning [9], efficient solutions for large volume of data are challenging. Thus, in order to be "smart," the data still needs to be cleaned even after its integration, and data preprocessing is the set of techniques utilized to encompass this task [10, 11].

An alternative way to deal with redundant or contradictory data is the use of data reduction techniques. These methods aim to reduce the original data, trying to maintain the integrity and information as much as possible. Their application is especially suitable when execution times of learning algorithms are prohibitive, forcing the practitioner to reduce the input size to a manageable volume with the maximum quality possible. In this area the most relevant reduction techniques

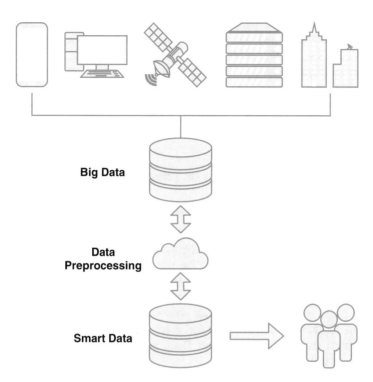

Fig. 3.1 Moving from "Big" to "Smart" Data

are feature selection (FS), instance selection (IS), and discretization algorithms. Some FS algorithms for Big Data have already been proposed [19, 22, 25] and IS techniques are also drawing the attention of researchers as well [29]. Discretization, which is a mandatory step for some learning algorithms, has been already tackled by adapting sequential procedures to modern Big Data frameworks [21].

We must also pay attention to the label distribution in supervised problems in Big Data problems. Class imbalance, which is already a challenge in classic machine learning problems, acquires a new dimension in Big Data, where the overwhelming amount of examples mislead learning algorithms to mostly consider only the majority examples. In traditional machine learning, data resampling was the preferred choice. While data resampling is already available in Big Data frameworks [5, 23], the introduced artificial minority examples can be a problem, as they increment the computing cost of the posterior learning algorithms. For this reason, novel preprocessing approaches are being explored by researchers [8].

Once the data is "smart," it can hold the valuable data and allows interactions in "real time," like transactional activities and other business intelligence applications. The goal is to evolve from a data-centered organization to a learning organization, where the focus is set on the knowledge extracted instead of struggling with the data management [13]. However, Big Data generates great challenges to achieve this since its high dimensionality and large example size imply noise accumulation, algorithmic instability and the massive sample pool is often aggregated from heterogeneous sources [7]. While FS, discretization or imbalanced algorithms to cope with the high dimensionality have drawn the attention of current Big Data frameworks (such as Spark's MLlib [17]) and researchers [21, 25, 27], algorithms to clean noise are still a challenge. In summary, challenges are still present to fully operate a transition between Big Data to Smart Data [28].

3.3   Smart Data and Internet of Things: Smart Cities and Beyond

More and more sensor-based devices are deployed for various applications to collect a huge amount of data for understanding the world. We have already mentioned that such data gathering process is at the foundation of Big Data. Traditionally, the connected devices would transfer all the monitored information to the cloud, under the cloud computing paradigm [1]. Although elastic to computational problem demands, cloud computing as a service presents latency problems. Extracting the value from data in these cases implies transferring and processing the huge volumes of raw data, transforming it to Smart Data, and sending back the decisions or knowledge obtained.

As the cost of the sensor devices drops, the cloud computing paradigm is evolving. Sensor's capabilities are increasing, enabling them to become smarter because they can generate Smart Data (which means processing out the noise and hold the valuable data) [26] at the very edge of the process. Internet of Things (IoT) [2] and the success of rich cloud services have pushed the horizon of a new computing paradigm, edge computing, which calls for processing the data at the edge of the network. Figure 3.2 depicts the displacement of computing task from the cloud servers to the very end of the connected, processing network. Thus, opposed to cloud computing the edge computing paradigm arises [24], where the processing (or at least part of it) is carried out locally, enabling lower response times and less network congestion.

The natural application of the edge computing power is to process the data, generating Smart Data locally, which can be rapidly exploited by stakeholders. Envisioning a technological landscape where most of the connected devices are smart in our everyday life has led to the term smart cities [30]. In such a vision, the unification of services is achieved by the local processing in IoT and by generating Smart Data in situ.

Nevertheless, smart cities are not the only scenario where smart devices will transform the society. Another prominent example is the presence of cyber-physical devices in industry, the so-called Industry 4.0 [14], in which information is closely monitored and synchronized between the physical factory floor and the cyber

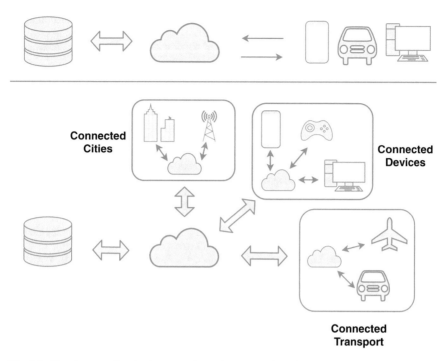

Fig. 3.2 Cloud computing (top) opposed to edge computing (bottom). The computing power moves from the remote servers to the network's edges, allowing lower latencies and new opportunities

computational space. Thanks to information analytics applied over the networked machines, the smart information processed and generated will enable the industry devices to perform more efficiently, collaboratively, and resiliently [18].

References

1. Armbrust, M., Fox, A., Griffith, R., Joseph, A. D., Katz, R., Konwinski, A., et al. (2010). A view of cloud computing. *Communications of the ACM, 53*(4), 50–58.
2. Atzori, L., Iera, A., & Morabito, G. (2010). The internet of things: A survey. *Computer Networks, 54*(15), 2787–2805.
3. Baldassarre, M. T., Caballero, I., Caivano, D., Rivas Garcia, B., & Piattini, M. (2018). From big data to smart data: A data quality perspective. In *Proceedings of the 1st ACM SIGSOFT International Workshop on Ensemble-Based Software Engineering* (pp. 19–24). New York: ACM.
4. Chen, J., Dosyn, D., Lytvyn, V., & Sachenko, A. (2017). Smart data integration by goal driven ontology learning. In *Advances in Intelligent Systems and Computing* (vol. 529, pp. 283–292).
5. del Río, S., López, V., Benítez, J. M., & Herrera, F. (2014). On the use of MapReduce for imbalanced big data using random forest. *Information Sciences, 285*, 112–137.

6. Fan, J., & Fan, Y. (2008). High dimensional classification using features annealed independence rules. *Annals of Statistics, 36*(6), 2605–2637.
7. Fan, J., Han, F., & Liu, H. (2014). Challenges of big data analysis. *National Science Review, 1*(2), 293–314.
8. Fernández, A., del Río, S., Chawla, N. V., & Herrera, F. (2017). An insight into imbalanced big data classification: outcomes and challenges. *Complex & Intelligent Systems, 3*(2), 105–120.
9. Frénay, B., & Verleysen, M. (2014). Classification in the presence of label noise: A survey. *IEEE Transactions on Neural Networks and Learning Systems, 25*(5), 845–869.
10. García, S., Luengo, J., & Herrera, F. (2015). *Data preprocessing in data mining.* Berlin: Springer.
11. García, S., Luengo, J., & Herrera, F. (2016). Tutorial on practical tips of the most influential data preprocessing algorithms in data mining. *Knowledge-Based Systems, 98*, 1–29.
12. García-Gil, D., Luengo, J., García, S., & Herrera, F. (2019). Enabling smart data: Noise filtering in big data classification. *Information Sciences, 479*, 135–152.
13. Iafrate, F. (2014). A journey from big data to smart data. *Advances in Intelligent Systems and Computing, 261*, 25–33.
14. Lee, J., Bagheri, B., & Kao, H.-A. (2015). A cyber-physical systems architecture for industry 4.0-based manufacturing systems. *Manufacturing Letters, 3*, 18–23.
15. Lenk, A., Bonorden, L., Hellmanns, A., Roedder, N., & Jaehnichen, S. (2015). Towards a taxonomy of standards in smart data. In *Proceedings: 2015 IEEE International Conference on Big Data, IEEE Big Data 2015* (pp. 1749–1754).
16. Marr, B. (2015). *Why only one of the 5 Vs of big data really matters.* https://www.ibmbigdatahub.com/blog/why-only-one-5-vs-big-data-really-matters/. Online; accessed July 2019.
17. Meng, X., Bradley, J., Yavuz, B., Sparks, E., Venkataraman, S., Liu, D., et al. (2016). MLlib: Machine learning in apache spark. *Journal of Machine Learning Research, 17*(34), 1–7.
18. Monostori, L., Kádár, B., Bauernhansl, T., Kondoh, S., Kumara, S., Reinhart, G., et al. (2016). Cyber-physical systems in manufacturing. *CIRP Annals, 65*(2), 621–641.
19. Peralta, D., del Río, S., Ramírez-Gallego, S., Triguero, I., Benitez, J. M., & Herrera, F. (2016). Evolutionary feature selection for big data classification: A MapReduce approach. *Mathematical Problems in Engineering, 2015*, 1–11, Article ID 246139
20. Raja, P. V., Sivasankar, E., & Pitchiah, R. (2015). Framework for smart health: Toward connected data from big data. *Advances in Intelligent Systems and Computing, 343*, 423–433.
21. Ramírez-Gallego, S., García, S., Mouriño-Talín, H., Martínez-Rego, D., Bolón-Canedo, V., Alonso-Betanzos, A., et al. (2016). Data discretization: taxonomy and big data challenge. *Wiley Interdisciplinary Reviews: Data Mining and Knowledge Discovery, 6*(1), 5–21.
22. Ramírez-Gallego, S., Lastra, I., Martínez-Rego, D., Bolón-Canedo, V., Benítez, J. M., Herrera, F., et al. (2017). Fast-mRMR: Fast minimum redundancy maximum relevance algorithm for high-dimensional big data. *International Journal of Intelligent Systems, 32*(2), 134–152.
23. Rastogi, A. K., Narang, N., & Siddiqui, Z. A. (2018). Imbalanced big data classification: A distributed implementation of smote. In *Proceedings of the Workshop Program of the 19th International Conference on Distributed Computing and Networking* (p. 14). New York: ACM.
24. Shi, W., Cao, J., Zhang, Q., Li, Y., & Xu, L. (2016). Edge computing: Vision and challenges. *IEEE Internet of Things Journal, 3*(5), 637–646.
25. Tan, M., Tsang, I. W., & Wang, L. (2014). Towards ultrahigh dimensional feature selection for big data. *Journal of Machine Learning Research, 15*, 1371–1429.
26. Teng, H., Liu, Y., Liu, A., Xiong, N. N., Cai, Z., Wang, T., et al. (2019). A novel code data dissemination scheme for internet of things through mobile vehicle of smart cities. *Future Generation Computer Systems, 94*, 351–367.
27. Triguero, I., del Río, S., López, V., Bacardit, J., Benítez, J. M., & Herrera, F. (2015). ROSEFW-RF: the winner algorithm for the ECBDL14 big data competition: an extremely imbalanced big data bioinformatics problem. *Knowledge-Based Systems, 87*, 69–79.

28. Triguero, I., García-Gil, D., Maillo, J., Luengo, J., García, S., & Herrera, F. (2019). Transforming big data into smart data: An insight on the use of the k-nearest neighbors algorithm to obtain quality data. *Wiley Interdisciplinary Reviews: Data Mining and Knowledge Discovery, 9*(2), e1289.
29. Triguero, I., Peralta, D., Bacardit, J., García, S., & Herrera, F. (2015). MRPR: A MapReduce solution for prototype reduction in big data classification. *Neurocomputing, 150*, 331–345.
30. Zanella, A., Bui, N., Castellani, A., Vangelista, L., & Zorzi, M. (2014). Internet of things for smart cities. *IEEE Internet of Things Journal, 1*(1), 22–32.

Chapter 4
Dimensionality Reduction for Big Data

4.1 Introduction

Popular 5Vs scheme has served to practitioners and data scientists to establish the most relevant challenges of Big Data [30]. Although there is a growing consensus about Big Data and the immediate challenges behind each "V," 5Vs scheme can be deemed as subjective and in continuous evolution. For example, when we talk about volume the idea of millions of tuples comes to our mind: several clients, patients, etc. However, it is noteworthy to remark that recent research on volume has only geared towards one side of the coin (longitude), and has obviated the "Big Dimensionality" side (broadness). Yet several works on high-dimensional small sample size problems are present in the literature, the study of high volume in both sides remains under-explored. The explosion of dimensions in real-world problem brings about new challenges to data analytics.

Causes of feature explosion are many and diverse. One is related to the different representations that data can take. As there exist many ways to describe a problem, there will be many alike representations, and thus, features to model it. Depending on expert's experience, background, and understanding of the input domain, feature descriptors can vary in order to meet the mental representation imposed by the expert. For instance, in text mining, practitioners work simultaneously with several feature types such as words, n-grams, part-of-speech tagging templates, etc., in order to achieve complete and comprehensive representations [26]. Not only text, but also imagery data with millions of features are common in today's real-world applications [18]. Finally, novel advancements in computing infrastructure, and outstanding technologies to process, storage and transmit distributed data have also boosted the development of Big Dimensionality.

Notice that in this context more features may not necessarily imply more discriminative power for algorithms. Features may just bring noise, or tiny importance to the prediction process. This fact motivates the application of some smart election among the overwhelming set of features so that only a reduced set of relevant features are

© Springer Nature Switzerland AG 2020
J. Luengo et al., *Big Data Preprocessing*,
https://doi.org/10.1007/978-3-030-39105-8_4

considered during the learning phase. Feature selection (FS) and feature extraction are among the most relevant tasks in data mining to accomplish this though task. Both schemes enable models to perform faster, reduce storage requirements, and in some cases improve accuracy (when reducing variance does not suppose an increment on biased error).

Nowadays, the current volume of data managed by our systems has surpassed the processing capability of standard systems and algorithms. Feature reduction is presented as an enabling remedy for large-scale learners. However, reduction methods first need to address the issues that prevent its operation on Big Data. Particularly, one has to cope with the explosive combinatorial effects of "curse of Big Dimensionality" while promoting high-value feature subsets from the original set of irrelevant, and redundant features.

In the rest of this chapter, we shall start with the discussion of the big dimensionality problem, as well as the benefits that can be exploited from this curse (Sect. 4.2). Afterwards we shall outline the proposals developed until now to deal with dimensionality reduction (Sect. 4.3). Hereinafter, we shall study in depth one of the most relevant distributed selection solutions for high-dimensional data (Sect. 4.4). Then, we study the problem of dimensionality reduction in Big Data streaming scenarios (Sect. 4.5). Finally, we summarize the chapter and draw some conclusions (Sect. 4.6).

4.2   The Curse of Big Dimensionality

Advancements on technology is one of the main causes of the development of Big Dimensionality. Existing gaps between contemporary processing and storage capacities that demonstrate our ability to capture and store data have far outpaced our ability to process and utilize it. Moore's law reports that processing capacity double every 18 months, while disk storage capacity doubles every 9 months (storage law).

For instance, technology of embedded cameras in cell phones is progressing steeply each year boosted by innovations in integrated hardware (flashlight, processor, sensor size, and techniques). The recent developments in smartphones and also in its image capturing and processing abilities are encouraging users to replace old-fashioned cameras by unique portable devices capable of capturing HD images and video. Among with these developments, the giant popularity of online social networks, like Facebook, Twitter, or Instagram, forecasts an accelerated expansion of phone pictures on the Internet.

Behind images, pixel is the basic unit adopted by most feature generators, and processing algorithms. Over the last decade, the resolution of embedded cameras in cell phones has grown from 0.11 megapixels (SHARP J-SH04)[1] to 4096 megapixels

---

[1] https://en.wikipedia.org/wiki/Camera_phone.

with new 4K-ready devices in 2017. Previous phenomenon evidences that the growth of dimensions is unstoppable right now, and seems to continue in next years, which will intensify the challenges in Big Data.

Another source of interest is the Internet. The content stored in the big web grows every day, and seems to be endless thanks to storage developments, among other technology advancements. Only the multimedia content stored in the Internet (text, image, 3D graphics, audio, and video) represents over 60% of its traffic [36], and online video more than half of multimedia content.

Online video provides even more descriptors (motion, acoustic, text, etc.) than pictures as video is far more complex than a sequence of images. In recent years video format has evolved with increasing resolution and metadata, reaching 4096 megapixels in each frame.

On the other hand, processing capabilities in today's systems have improved too in order to give support to the increasing amount of users browsing through the Internet. Uploading and presenting videos, pictures, and other content are some activities that demand large-scale analysis, summarization, collection, aggregation, and querying.

Feature uptrend is not only specific from the multimedia field, but also can be found in science. For instance, microarray data consists of thousands of biomarkers (features) where only some relevant genes are selected for further *in vitro* study [15]. This task is usually performed by an analytical program which discerns between relevant and useless genes. Thanks to the recent developments in biotechnology, new bio-inspired formats have arisen to more concisely define the behavior of genes. Single-nucleotide polymorphism is a novel format that allows representations of millions of features, which is a quantum leap compared with the thousand scale used previously.

Focusing on popular public datasets, we perform an analysis on the rise of Big Dimensionality in two popular data science repositories, namely UC Irvine Machine Learning Repository (UCI) [16] and LIBSVM [9]. UCI and LIBSVM have evolved from small samples to larger sizes up to millions of dimensions. LIBSVM covers a wide spectrum of datasets from many fields such as imagery (Corel), biology (Leukemia), physic (Higgs), text (Webspam), time-series (Gas Sensor), and video (YouTube MVG), among others. Table 4.1 outlines the main characteristics of some datasets created in the period 1985–2015, and uploaded to these repositories.

To illustrate the evolution of dimensions on these popular repositories, we depict in Fig. 4.1 the number of features for datasets published in these repositories from the 1985 to 2015. From here, an exponential growth in dimensionality can be highlighted across all repositories. For example, in LIBSVM, the smallest dataset is USPS (256 dimensions), whereas the largest one is KDD2010 (almost 30 million dimensions). This means that in 2010 the upper limit in size is 100,000 times larger than in only 16 years before. Following this upward trend, we may assert that future systems will need to be prepared to deal with billions of features.

Worst of all, Big Dimensionality is far from being embraced by current research in FS. In [47], authors show that the dimensionality managed by a long list of algorithms lags significantly behind that produced by current datasets. As a further

Table 4.1 Evolution of dimensions (# features) in popular data repositories (UCI and LIBSVM) from 1985 to 2015

Repository	Dataset	# Dimensions	Year
LETTER	UCI	16	1991
SOYBEAN	UCI	35	1987
CHESS	UCI	36	1989
H2O	UCI	38	1993
CONNECT-4	UCI	42	1995
SPECTF	UCI	44	2001
MOLECULAR	UCI	58	1990
MAMMALS	UCI	72	1992
INSURANCE	UCI	86	2000
COREL	UCI	89	1999
SPECTROMETER	UCI	102	1988
ISOLET	UCI	617	1994
INTERNET AD	UCI	1558	1998
P53	UCI	5409	2010
BAG OF WORDS	UCI	100,000	2008
PEMS-SF	UCI	138,672	2011
YouTube MVG	UCI	1,000,000	2013
Gas sensor	UCI	1,950,000	2013
URL	UCI	3,231,961	2009
USPS	LIBSVM	256	1994
COLON-CANCER	LIBSVM	2000	1999
GISETTE	LIBSVM	5000	2003
Leukemia	LIBSVM	7129	1999
BREAST-CANCER	LIBSVM	7129	2001
REAL-SIM	LIBSVM	20,958	1998
SIAM	LIBSVM	30,438	2007
RCV1	LIBSVM	47,236	2004
SECTOR	LIBSVM	55,197	1998
News20	LIBSVM	62,061	1995
News20.binary	LIBSVM	1,355,191	2005
LOG1P	LIBSVM	4,272,227	2009
Webspam	LIBSVM	16,609,143	2006
KDD2010	LIBSVM	29,890,095	2010

proof, authors show that less than 80% of methods do not dare to process datasets with more than 10,000 dimensions, and only a 5% of these methods tackle datasets with millions of features.

Given this gap between immense dimensionality and outdated methods, it has surged an urgent need for creating novel approaches, paradigms, and methodologies that can deal with such amount of features. Novel dimensionality reduction is required to cope with complex requirements, such as fast processing and reduced storage, while providing relevant subset of features. Large-scale feature reduction can be addressed from two perspectives: either by designing novel sequential

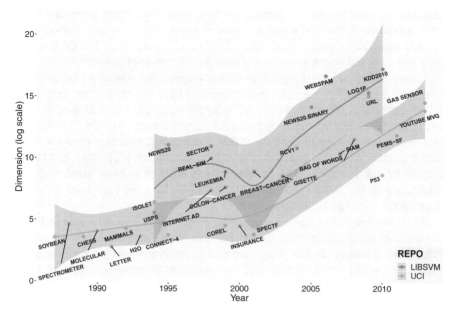

Fig. 4.1 Evolution of dimensions (# features) in common datasets from popular repositories (UCI and LIBSVM) from 1985 to 2015

methods that explicitly address the requirements above or by creating distributed solutions inspired by standard methods [35]. Neither of them are easy to implement as several technical and algorithmic nuances must be carefully addressed.

Focusing on sequential approaches, filter methods [6] (rank features by some criteria) could be deemed as a sure bet in the Big Data context because of its great efficiency, and the possibility of offering preliminary results throughout a greedy selection process. Nevertheless, most of the filters mainly focus on predictive information and pay little attention on feature relations. This behavior negatively affects problems where an important number of intrinsic interactions are present, such as in microarray data analysis. Furthermore, filter algorithms usually get trapped in local optima.

A possible solution here is to replace individual scores by some kind of subset-level contribution measurement, which evaluate the importance of linkages between selected features. Some examples that implement inter-feature interactions are graph-based FS algorithms [29], and some information-based selectors like minimum redundancy maximum relevance (mRMR) [32]. The latter approach, for instance, selects those features more related to the class, and less related among them.

Although more effective, subset-based selectors fail whenever they have to compute interactions between millions of dimensions since its underlying computational time complexity is quadratic. Imagine the case of KDD2010, where millions of comparisons must be performed between features (pairwise comparison). This originates trillions of pairwise correlations to be computed, which is intractable for current systems. Generally, many empirical works have shown that filter selectors are not designed to scale above tens of thousands of features. Notice that this supposes a big and unprecedented challenge for the data science community.

Some notable wrapper (rely on classifier weights) and embedded (selection is integrated in learning) methods in the literature have proven to cope with huge set of dimensions (from thousands of features) elegantly. For example, support vector machines (SVM) based on recursive feature elimination (SVM-RFE) [19] rely on support vectors to evaluate feature individually, and then removes those less important from the final set. For a leap in efficiency, l_1-norm regularizer may be included in SVM [17]. In recent years, other techniques based on group discovery [48] and feature generation machines [40] have been proposed to further improve selection in large environments.

As mentioned before feature correlation possesses an important role in FS. It is especially important in methods that evaluate groups of features. Graph-guided fused lasso is among the first group selection techniques in using a graph to select groups. Octagonal shrinkage and clustering algorithm for regression reduce similar feature tuples by adding the l_∞-regularizer [7]. An efficient projection strategy was introduced in [50] to boost up the grouping process.

Besides the drawbacks previously presented, there still exist good news about Big Dimensionality (the blessing of Big Dimensionality), though they are less abundant. A correlation study performed by Zhai et al. [48] on two versions of the News20 dataset shows that more than 99% of features present correlation scores lower than 0.1. This comes to say that most of the features are either uncorrelated or those correlated are extremely sparse. In fact, correlated features are far less abundant in News20.binary (1,355,191 dimensions) than in News20 (62,601 dimensions). Previous results are quite positive in the sense that FS could leverage upon the inversely proportional relationship between dimensionality and correlation to easily select a very reduced group of relevant features.

4.3   Distributed Proposals for Dimensionality Reduction

This section aims at detailing a thorough list of distributed contributions on Big Data dimensionality reduction. Table 4.2 classifies all contributions in the literature according to the following features: number of features, number of instances, maximum size managed by each algorithm, as well as the framework upon they have been developed. The size has been computed by multiplying the total number features by the number of instances (8 bytes per datum). For sparse methods (e.g.: [37] or [40]), only the non-sparse cells have been considered.[2]

---

[2]Although feature generation machine is not a distributed method as such, it has been included here for its outstanding relevance in the comparison.

Table 4.2 Dimensionality reduction methods for Big Data

Methods	# Features	# Instances	Size (GB)	Framework
[35]	630	65,003,913	305.1196	Apache Spark
[33]	630	65,003,913	305.1196	Hadoop MapReduce
[42]	630	65,003,913	305.1196	Hadoop MapReduce
[49]	1156	5,670,000	48.8350	MPI
[40]	29,890,095	19,264,097	4.1623	C++/MATLAB
[37]	100,000	10,000,000	1.4901	MapReduce
[31]	100	1,600,000	1.1921	Apache Spark
[43]	127	1,131,571	1.0707	Hadoop MapReduce
[23]	54,675	2096	0.8538	Hadoop MapReduce
[39]	54	581,012	0.2338	Hadoop MapReduce
[11]	20	1,000,000	0.1490	MapReduce
[10]	–	–	0.0976	Hadoop MapReduce
[20]	256	38,232	0.0729	Hadoop MapReduce
[46]	11,852	50,000	0.0397	Hadoop MapReduce
[44]	20,000	2600	0.0387	Hadoop MapReduce
[25]	2728	13,000	0.2642	Twister
[41]	52	5253	0.0020	Hadoop MapReduce
[13]	–	–	0.0000	Hadoop MapReduce
[27]	–	–	0.0000	Hadoop MapReduce
[21]	–	–	0.0000	Hadoop MapReduce

The methods are ordered by total size (number of features × number of instances × 8 bytes/datum). Those methods with no information about number of features or instances have been set to zero

From Table 4.2, we can conclude that methods in [33, 35, 42] are the only capable of selecting features in datasets with hundreds of gigabytes, whereas [35, 40] are capable of dealing with millions of features. From the previous list, we can highlight the algorithm from [35] as the most competitive option since it is able to process huge datasets in both dimensions (feature and instance side). It is also noteworthy to remark that most of the methods are developed on Hadoop MapReduce; a platform that has largely shown its deficiencies at dealing with learning/iterative processes. Additionally, most of the methods test their performance on datasets with less than a thousand of features, which is quite far from reality as shown in Fig. 4.1.

As mentioned before, FS has a key role to play in dealing with large-scale datasets, especially those that present an ultra-high dimensionality. However, FS methods, like many other learning methods, suffer from the "curse of dimensionality" [47], and consequently, are not expected to scale well. New paradigms and tools have emerged to give support to this task [5], such as Apache Spark, Flink, or Hadoop MapReduce. Most of them are centered in the use of parallel processing to distribute the massive complexity burden across several nodes. Here, a list of the contributions for FS for Big Data is presented:

- [35]: authors designed a novel distributed FS framework for the Apache Spark platform. It was inspired by a previous sequential framework based on information theory and developed by Brown et al. [8]. The distributed framework includes most of the state-of-the-art filtering selectors in the literature, like mRMR or InfoGain. By considering redundancy and relevance measures in their computations, this framework is able to evaluate interactions of features from two perspectives: redundancy and relevance of features. This model will be described in detail in Sect. 4.4 as an illustrative example of how to distribute the FS process in a cluster environment.

- [33]: Peralta et al. proposed a different approach (called MR-EFS) based on independent genetic algorithm processes (executed on each partition), and a voting scheme to aggregate partial solutions.

  By using a single MapReduce phase, MR-EFS is able to select the most relevant features in the input dataset. The map phase in MR-EFS consists of applying a CHC-based FS algorithm on each data partition. The result is a binary vector indicating if each feature is locally selected or not. Then, a single reduce phase is performed by averaging partial binary vectors. The final scheme is sent to the master for the final reduction phase. Dataset reduction is achieved by applying an extra MapReduce phase. Each map processes an instance and generates a new one that only contains the features previously selected. Finally, the instances generated are concatenated to form the final reduced dataset, without the need for an extra reduce phase.

- [42]: Triguero et al. proposed an evolutionary feature weighting model to learn the feature weights per map. They introduced a reduce phase adding the weights and using a threshold to select the most relevant instance. This model was the winner solution in the ECBDL'14 competition.

  In this algorithm, each Map performs a whole differential evolution feature weighting cycle in order to weight features according to the local structure of data. That is, a complete loop of mutation, crossover, and selection operators for a given number of iterations. Then, mappers will emit a resulting vector of weights measuring the importance of each feature regarding this subset of the training set. The proposed scheme only uses one single reducer to sum all the partial weights in each map, so that aggregated information by feature is provided to the master node.

- [49]: Zhao et al. proposed a FS framework for both unsupervised and supervised learning, which includes several measures, such as the Akaike information criterion, the Bayesian information criterion, and the corrected Hannan–Quinn information criterion. This framework has been implemented on message passing interface (MPI).

- [40]: Tan et al. propose an adaptive feature scaling scheme that reformulates the FS problem as a convex semi-infinite programming (SIP) problem. They first control the number of features to be selected by scaling and constraining (through l_1-norm) input features. The resulting problems are then transformed to meet the convex requirement by applying a convex SIP scheme. A feature generation machine is chosen as solver for this transformed problem, which includes top relevant features per iteration and solves a sequence of much reduced multiple kernel learning subproblems. Authors certify the global convergence of the model.

In order to improve the scalability of the method, Tan proposes to address the primal form of the multiple kernel problems through a modified approximate primal gradient method. The utilization of cache is also attached to further improve the performance.

One of the main advantages of feature generation machines is that as they only work with a small subset of features (kernels) in the subproblem optimization, they are particularly appropriate for high-dimensional problems.

- [37]: Singh et al. proposed a new approximate heuristic, optimized for logistic regression in MapReduce, which employs a greedy search to select features increasingly. It also utilizes approximate log-likelihood computations based on histogram data to speed up the whole process.
- [31]: A filter method based on column subset selection was implemented by Ordozgoiti et al. However, as stated by the authors, this Spark algorithm is designed for datasets with millions of features, but not for high-dimensional problems.
- [43]: Wang et al. designed a family of FS algorithms for online learning. The algorithm selects those features with larger weights, according to a linear classifier based on L1-norm.
- [23]: Kumar et al. implemented three FS algorithms (ANOVA, Kruskal–Wallis, and Friedman test) based on statistical test. All of them were parallelized on Hadoop MapReduce so as each feature is evaluated independently.
- [39]: Sun et al. designed a method that computes the total combinatory mutual information, and the contribution degree between all feature variables and class variable. It uses an iterative process (implemented on Hadoop) to select the most relevant features.
- [11]: A FS method based on differential privacy (Laplacian Noise) and a Gini-index measure was designed by Chen et al. This technique was implemented using a general MapReduce model. In the map phase, the algorithm computes the Gini value for each feature on each data partition. Then, these values are aggregated to form a rank of features.
- [10]: A simple version of TF-IDF (for Hadoop MapReduce) was designed by Chao et al. to deal with text mining problem on Big Data. In [25], another approach for FS in text mining is presented. In each map, TF-IDF information is used to compute the probability distribution function for each feature, assuming that all of them are Gaussian. In the same map, mutual information scores are calculated and sent to the reduce phase, which compares and selects the best ones.
- [20]: He et al. implemented on Hadoop a FS method using positive approximation as an accelerator for traditional rough sets.
- [46]: Eftim et al. propose an information-based FS algorithm, similar to [35]. In counterpart to [35], which greatly reduces the number of pairs generated, this framework computes information gain by shuffling all pairs feature-value

and class across the network. Although not very elegant, this process allows to calculate simple and conditional likelihoods for the subsequent selection phase.

- [44]: A parallel implementation of the Relief algorithm is proposed here. Relief is based on the idea that relevant features are those that help to distinguish close instances in the input space. Returning to the distributed design, it performs vertical splitting of original data to create subsets of features distributed across the cluster. In the map phase, the Relief algorithm is applied on each subset to classify the features by a weighting measure. These weights will serve to select those features with a scores greater than a given threshold (reduce phase).
- [41]: Tanupabrungsun et al. proposed a genetic algorithm approach with a wrapper fitness function. K-nearest neighbors are used to evaluate the solutions in binary format (one gene per feature). In this work, the Hadoop master process is in charge of the management of the population, whereas the fitness evaluation is parallelized. To do that the whole test set is replicated across all nodes so that the computation of distances is also parallel. The reducer combines solutions to obtain the generational results.
- [13]: Dalavi et al. proposed a novel weighting scheme based on supervised learning (using SVMs) for Hadoop MapReduce. Data nodes are responsible for converting the documents into contribution vectors, which will be evaluated by a SVM classifier.
- [27]: Meena et al. designed an evolutionary approach based on ant colony optimization with the aim of finding the optimal subset of features. It parallelizes on Hadoop MapReduce some parts of the algorithm, such as tokenization, the computation of association degrees, and the evaluation of solutions.
- [21]: Hodge et al. proposed an unified framework which uses binary correlation matrix memories to store and retrieve patterns using matrix calculus. They propose to compute sequentially these matrices, and then, to distribute them on Hadoop to obtain the final coefficients.

Concluding this section, we remark that almost every method in this list evaluates features independently without considering complex interactions between features (direct parallelization). Only methods such as [33, 35, 40, 42] consider the underlying relationships between the whole set of features. Although more complex to implement, we want to remark that inclusion of feature interactions is crucial to improve the performance of selection algorithms in this new context of Big Data.

4.4   An Information Theoretical Feature Selection Framework for Apache Spark

In this section, we will analyze an information theoretical FS framework, focused on the large-scale FS problem [35]. This algorithm is inspired by an information-based FS framework proposed by Brown et al. [8], which includes many common filtering methods. In this work, Brown et al. show that algorithms such as mRMR and other

SparkPackages `spark-infotheoretic-feature-selection`

This package contains a generic implementation of greedy Information Theoretic Feature Selection (FS) methods. The implementation is based on the common theoretic framework presented by Gavin Brown. Implementations of mRMR, InfoGain, JMI and other commonly used FS filters are provided.

```
spark-shell --packages sramirez:spark-infotheoretic-
feature-selection:1.4.4
```

http://spark-packages.org/package/sramirez/
spark-infotheoretic-feature-selection

Fig. 4.2 Spark package: information theoretical FS framework

algorithms are special cases of conditional mutual information (CMI) when certain specific independence assumptions are made about both the class and input features. The distributed approach designed in [35] proves that these criteria are not only a sound theoretical formulation, but also fit well within modern Big Data platforms, and allow us to distribute several FS methods and their complexity across a cluster of machines.

This framework contains a generic implementation of several information theory-based FS methods—including mRMR, conditional mutual information maximization (CMIM), and joint mutual information (JMI). In Fig. 4.2 we can find a Spark Package associated with this research.

Below we first briefly present Brown's framework, and analyze its adaptation to the Big Data environment. We then describe in detail how the selection process and the underlying information theory operations were implemented in a distributed manner using Spark's primitives.

4.4.1 Information Theory-Based Filter Methods

Information measures tell us how much information has been acquired by the receiver when sent a message [28]. In predictive learning, we associate the message with the output feature in classification.

A commonly used uncertainty function is mutual information (MI) [12], which measures the amount of information one random variable contains about another, in other words, it expresses the reduction in the uncertainty of one random variable due to knowledge of the other variable:

$$I(A; B) = H(A) - H(A|B)$$

$$= \sum_{a \in A} \sum_{b \in B} p(a, b) \log \frac{p(a, b)}{p(a)p(b)}, \qquad (4.1)$$

where A and B are the two random variables with marginal probability mass functions $p(a)$ and $p(b)$, respectively, $p(a, b)$ is the joint mass function, and H is the entropy.

MI can likewise be conditioned to a third random variable. Thus, CMI is denoted as:

$$I(A; B|C) = H(A|C) - H(A|B, C)$$

$$= \sum_{c \in C} p(c) \sum_{a \in A} \sum_{b \in B} p(a, b, c) \log \frac{p(a, b, c)}{p(a, c)p(b, c)}, \qquad (4.2)$$

where C is a third random variable with marginal probability mass function $p(c)$ and $p(a, c)$, $p(b, c)$, and $p(a, b, c)$ are the joint mass functions.

Filtering methods are based on a quantitative criterion or index, also known as the relevance index or scoring. This index measures the usefulness of each feature for a specific classification problem. Through the **relevance** of a feature for the class (self-interaction), we can rank features and select the most relevant ones. However, features can also be ranked using a more complex criterion such as whether it is more **redundant** than another feature (multi-interaction). For instance, redundant features (variables that carry similar information) can be discarded using the MI criterion [4]:

$$J_{mifs}(X_i) = I(X_i; Y) - \beta \sum_{X_j \in S} I(X_i; X_j),$$

where $S \subseteq S_\theta$ is the current set of selected features and β is a weight factor. Considered is the MI between each candidate $X_i \notin S$ and the class. Also introduced is a penalty proportional to the redundancy, calculated as the MI between the current set of selected features and each candidate feature.

A wide range of methods have been described in the literature that are built on these information theory measures. To homogenize the use of all the criteria, Brown et al. [8] proposed a generic expression that allows multiple information theory criteria to be ensembled in a single FS framework, based on a greedy optimization process that assesses features using a simple scoring criterion. Through certain independence assumptions, many criteria can be transformed as linear combinations of the Shannon entropy terms MI and CMI [12]. In some cases, more complex criteria are expressed as non-linear combinations of these terms (e.g., max or min). For a detailed description of the transformation processes and a comprehensive list of adapted FS methods, please refer to [8]. The generic formula proposed by Brown is as follows:

$$J = I(X_i; Y) - \beta \sum_{X_j \in S} I(X_j; X_i) + \gamma \sum_{X_j \in S} I(X_j; X_i | Y), \qquad (4.3)$$

where γ represents a weight factor for the conditional redundancy component.

The formula can be divided into three components, representing the relevance of a feature X_i, the redundancy between two features X_i and X_j, and the conditional redundancy between two features X_i, X_j and the class Y. Through the aforementioned assumptions, many criteria were re-written to fit the generic formulation in such a way that all the methods could be implemented using a slight variation on this formula.

4.4.2   Feature Selection Filtering Framework for Big Data

This section starts by describing the proposed FS framework for Big Data; outlining the main changes made to adapt the classical approach to the new Big Data environment. The implications arising from the distributed implementation of Eq. (4.3), as well as the complexity derived from the parallelization of core operations (namely MI and CMI) is also given.

The main improvements performed to make possible the redesign of Brown's framework are described below:

- **Columnar transformation:** The access pattern presented by most FS methods is feature-wise, in contrast to many other machine learning algorithms, which are instance-wise (operated by row). Although this may be considered a minor issue, it can significantly degrade the performance since the natural way to compute relevance and redundancy in FS methods is normally via columns. This issue is especially important for distributed frameworks like Spark, where the data partitioning scheme has a significant impact on performance.
- **Broadcasting features**: Once all features values have been grouped and partitioned into different partitions, data movement has to be minimal to avoid superfluous network and CPU usage. If the MI process is performed locally in each partition, the overall algorithm will run efficiently (almost linearly). Data movement is minimized by replicating the output feature and the last selected feature in each iteration.
- **Pre-computed data caching:** The first term in the generic criterion of Eq. (4.3) is relevance, which basically implies calculating MI for all the input features and the output (relevance). This operation is performed once at the start of our algorithm, then cached to be reused in subsequent evaluations of Eq. (4.3). Likewise, subsequent marginal and joint proportions derived from these operations are also kept so as to omit some computations. This also helps isolate redundancy computation by feature as it replicates permanent information in all nodes.

Algorithm 1 Main FS algorithm

Input: D Dataset, an RDD of samples.
Input: ns Number of features to select.
Input: $npart$ Number of partitions to set.
Input: $cindex$ Index of the output feature.
Output: S_θ Index list of selected features
 1: $D_c \leftarrow columnarTransformation(D, ns, npart)$
 2: $ni \leftarrow D.nrows; nf \leftarrow D.ncols$
 3: $REL \leftarrow computeRelevances(D_c, cindex, ni)$
 4: $CRIT \leftarrow initCriteria(REL)$
 5: $p_{best} \leftarrow CRIT.max$
 6: $sfeat \leftarrow Set(p_{best})$
 7: **while** $|S| < |S_\theta|$ **do**
 8:     $RED \leftarrow computeRedundancies(D_c, p_{best}.index)$
 9:     $CRIT \leftarrow updateCriteria(CRIT, RED)$
10:     $p_{best} \leftarrow CRIT.max$
11:     $sfeat \leftarrow addTo(p_{best}, sfeat)$
12: **end while**
13: **return** $(sfeat)$

- **Greedy approach:** Brown proposed a greedy search process in which only one feature is selected in each iteration. This approach transforms the quadratic complexity of typical FS algorithms into a more manageable complexity determined by the number of features to select.

Algorithm 1 is the main FS algorithm in this framework, in charge of deciding which feature to select in a sequential manner. Roughly speaking, it calculates the initial relevance for all the features, and then iterates to select the best features according to Eq. (4.3) and the underlying MI and CMI values.

The first step consists of transforming the data into columnar format. Algorithm 2 shows in detail this transformation. The idea behind this transformation is to transpose the local data matrix provided by each partition. The result is a new local matrix, so that each row contains the local points belonging to a given feature. In order to put all features parts together in the same data partitions, resulting rows are sorted by key (feature ID, data partition ID). This operation maintains the partitioning scheme and reduces the number of pairs sent to be shuffled. Additionally, once data are transformed, they can be cached and reused in the subsequent loop.

Figure 4.3 depicts this process using a small example with four instances in a single partition, and n features. This figure shows how all instances in a partition are transposed to form new rows/elements, one per feature. Then, a key-value pair is generated for each new row, where the key is formed by the feature id and the partition id, and the value by a single feature row. Finally, pairs with the same feature id are placed in the same partition, and locally ordered by old partition id.

Once the data matrix is transformed, the algorithm obtains the relevance for each feature in X, initializing the criterion value (partial result according to Eq. (4.3)) and creating the initial ranking of the features. Relevance values are saved as part of the

---

Algorithm 2 Function that transforms row-wise data into a column-wise format *(columnarTransformation)*

Input: D Dataset, an RDD of samples.
Input: nf Number of features.
Input: $npart$ Number of partitions to set.
Output: Column-wise data (RDD of feature vectors).
  1: $D_c \leftarrow$
  2: **map partitions** $part \in D$
  3:      $matrix \leftarrow new\ Matrix(nf)(part.length)$
  4:      **for** $j = 0$ *until* $part.length$ **do**
  5:         **for** $i = 0$ *until* nf **do**
  6:             $matrix(i)(j) \leftarrow part(j)(i)$
  7:         **end for**
  8:      **end for**
  9:      **for** $k = 0$ *until* nf **do**
 10:          $EMIT < k, (part.index, matrix(k)) >$
 11:      **end for**
 12: **end map**
 13: **return** $(D_c.sortByKey(npart))$

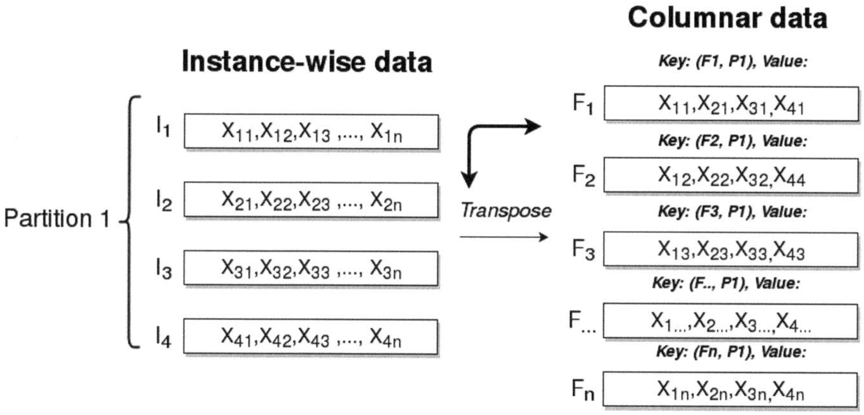

Fig. 4.3 Columnar transformation scheme. F and I indicate features and instances, respectively. Each rectangle on the left represents a single register in partition 1, and each rectangle on the right represents a transposed feature block in the new columnar format. A key-value pair is finally generated for each new row

previous expression and are reused in subsequent steps to update the criteria. The most relevant feature, p_{best}, is then selected and added to the set $sfeat$, initially empty. The iterative phase begins by calculating MI and CMI between p_{best}, each candidate X_i and Y. The resulting values update the accumulated redundancies (simple and conditional) of the criteria. In each iteration, the most relevant candidate features are selected as the new p_{best} and are added to $sfeat$. The loop ends once ns features (where $ns = |S_\theta|$) have been selected or when no more features remain to be selected.

Thanks to the columnar format, relevance and redundancy functions are able to compute scores for features in each partition locally. This is done by replicating/broadcasting[3] only the last selected feature and the class across the nodes, or just the class for relevance. As result, histograms for all the candidate features with respect to the auxiliary variables are computed locally.

As shown in Sect. 4.4.1, joint and marginal proportions needed to compute mutual information values can be easily calculated from histograms by aggregating rows and columns, and fetching individual cells from these tables. These likelihoods will serve to obtain the simple and conditional mutual information increments that update the ranking of features in each iteration. In case of relevance, plenty of information can be saved and reused in further iterations, for example, prior and posterior probabilities. Algorithm 3 depicts how relevances are computed in a distributed way, whereas Algorithm 4 focus on redundancy. Concerning redundancies, 3-dimensional histograms are generated involving an extra column that represents the last selected feature.

Algorithm 3 Compute mutual information between the set of features X and Y. *(computeRelevances)*

Input: D_c RDD of tuples (index, (block, vector)).
Input: $yind$ Index of Y.
Input: ni Number of instances.
Output: MI values for all input features.
1: $ycol \leftarrow D_c.lookup(yind)$
2: $bycol \leftarrow broadcast(ycol)$
3: $counter \leftarrow broadcast(getMaxByFeature(D_c))$
4: $H \leftarrow getHistograms(D_c, yind, bycol, null, null)$
5: $joint \leftarrow getProportions(H, ni)$
6: $marginal \leftarrow getProportions(aggregateByRow(joint), ni)$
7: **return** $(computeMutualInfo(H, yind, null))$

Algorithm 4 Compute CMI and MI between p_{best}, the set of candidate features, and Y. *(computeRedundancies)*

Input: D_c RDD of tuples (index, (block, vector)).
Input: $jind$ Index of p_{best}.
Output: CMI values for all input features.
1: $jcol \leftarrow D_c.lookup(jind)$
2: $bjcol \leftarrow broadcast(jcol)$
3: $H \leftarrow getHistograms(D_c, jind, bjcol, yind, bycol)$
4: **return** $(computeMutualInfo(H, jind, yind))$

---

[3]Broadcast operation in Spark sends a single copy of the variable to each node.

4.4.3 High-Dimensional Feature Selection: An Experimental Framework

In this section we show that the distributed framework presented above for large-scale selection perform well in both sides of Big Data—number of dimensions and longitude. Additionally, this section also aims at showing that FS also proves useful in Big Data classification, where most of the features tend to be irrelevant and have no influence on the output variable (Sect. 4.2).

For such purpose, the previous framework is applied on two different scenarios, one with millions of features and millions of sparse instances, and another with millions of instances, but a more manageable but still complex number of features (less than one thousand). The dataset elected for the first task is *ECBDL14*, used as a reference dataset at the international GECCO-2014 conference, and which consists of 631 features (including both numerical and categorical attributes) and 32 million instances. In this binary classification problem the class distribution is imbalanced, with 98% of negative instances. To fix the imbalance problem, we have applied the MapReduce version of the Random OverSampling (ROS) algorithm [14] to generate a final dataset with 65 millions of examples. The second dataset is *kddb*, a sparse dataset with almost 30 millions of features and 20 millions of training instances which come from the LibSVM dataset repository [9].

For the comparison study, the mRMR algorithm [32] has been chosen as the FS algorithm, SVM, and Naïve Bayes [1] as the reference classifiers (already implemented in the MLlib library [38]). The parameters of the classifiers were set as recommended in the authors' specifications. For evaluation purposes, two common evaluation metrics were used to evaluate the quality of selections: area under the receiver operating characteristic curve (AUROC, henceforth AUC) to evaluate classifier accuracy, and training modeling time to evaluate FS performance.

Finally, the maximum level of parallelism (number of partitions) was set to 864, twice the total number of cores available in the cluster. The cluster used is composed of eighteen computing nodes and one master node. The computing nodes have the following features: 2 processors x Intel Xeon CPU E5-2620, 6 cores per processor, 2.00 GHz, 15 MB cache, QDR InfiniBand Network (40 Gbps), 2 TB HDD, 64 GB RAM. We used the following configuration for the software: Hadoop 2.5.0-cdh5.3.1 from Cloudera's open-source Apache Hadoop distribution,[4] HDFS replication factor 2, HDFS default block size 128 MB, Apache Spark and MLlib 1.2.0, 432 cores (24 cores/node), 864 RAM GB (48 GB/node).

Time performance of the framework is evaluated in Table 4.3. It presents the time results obtained by mRMR using different ranking thresholds (number of selected features).

---

[4]http://www.cloudera.com/content/cloudera/en/documentation/cdh5/v5-0-0/CDH5-homepage. html.

Table 4.3 Selection time by dataset and threshold (in seconds)

# Features	kddb	ECBDL14
10	283.61	332.90
25	774.43	596.31
50	1365.82	1084.58
100	2789.55	2420.94

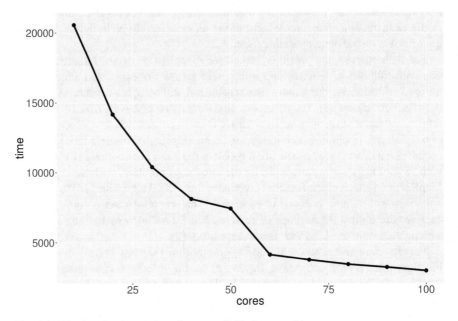

Fig. 4.4 Selection time by number of cores available (in seconds)

As can be observed here the distributed solution produces competitive results in every vase, irrespective of the number of iterations accomplished (represented by the number of features to select). It is noteworthy to remark that the model was able to rank 100 features in less than 1 h in both datasets.

As an extra study on scalability we have measure the effect of increasing the number of cores in the cluster. *ECBDL14* was utilized as reference with the same configuration. Figure 4.4 depicts how distributed selection performed depending on the number of cores (10–100). As expected, the study reveals a decreasing behavior as the number of cores increased.[5]

Usefulness of selection for large-scale classification is evaluated in Table 4.4. This show the accuracy results for SVM and Naive Bayes using different FS schemes on *ECBDL14*. The *kddb* dataset was omitted because none of the classifiers was able to run on the whole dataset.

---

[5]Note that the whole memory available in the cluster was only available from the 10-core value.

Table 4.4 AUC results for SVM and Naïve Bayes classification with reduced set of features

# Features	ECBDL14 (SVM)	ECBDL14 (NB)
All	0.5070	0.5122
10	0.5074	0.5144
25	0.5069	0.5155
50	0.5078	0.5148
100	0.5066	0.5153

Results show that there is no improvement or loss in accuracy when using FS. Although it is not the best scenario, reduction here is highly positive since subsequent learning phase will be surely more simple and rapid with far less features. Even with only ten features selected, AUC remains the same. Results also show that plenty of features do not provide extra information to this problem.

The thorough study performed in this section has shown that the distributed solution is capable of selecting features in a competitive time interval when applied to datasets that are huge—in both number of instances and features. Furthermore, solutions have shown to be useful on improving the simplicity of data and the subsequent learning process.

4.5   Dimensionality Reduction in Big Data Streaming

FS is one of the most extended data preprocessing techniques. Although we can find many proposals for static Big Data preprocessing, there is little research devoted to the continuous Big Data problem. Apache Flink is a recent and novel Big Data framework, following the MapReduce paradigm, focused on distributed stream and batch data processing [3].

In this section, we will describe a data stream library named DPASF (Data Preprocessing Algorithms for Streaming in Flink), focused on Big Data stream preprocessing [2]. The library is composed of six of the most popular and widely used data preprocessing algorithms for Apache Flink. It contains three algorithms for performing FS, and three algorithms for discretization. In this section, we focus on the FS methods. In Fig. 4.5 we can find the algorithm implementations associated with this research.

4.5.1   Information Gain

This FS scheme described in [22] is composed of two steps: an incremental feature ranking method, and an incremental learning algorithm that can consider a subset of the features during prediction.

FlinkML   DPASF

Big Data library oriented to online data preprocessing for Apache Flink. This library contains six of the most popular and widely used algorithms for data preprocessing in data streaming. It is composed of three feature selection algorithms and three discretization algorithms.

https://sci2s.ugr.es/BigDaPFlink

Fig. 4.5 FlinkML: DPASF library

Algorithm 5 InfoGain algorithm

Input: *data* a DataSet LabeledVector (label, features)
Input: *selectNF* Number of features to select
Output: DataSet with the most *selectNF* important features
1: $freqs \leftarrow frequencies(data, groupBy = label)$
2: $H \leftarrow Entropy(freqs)$
3: $gains \leftarrow$
4: **map** $i \in 0$ until $nFeatures$
5:      $freqs \leftarrow frequencies(data, feature_i)$
6:      $px \leftarrow probs(freqs)$
7:      $H \leftarrow entropy(freqs)$
8:      $H(Y|Feature_i) \leftarrow ConditionalEntropy(freqs)$
9:      $H - H(Y|Feature_i)$
10: **end map**
11: **return** $selectFeatures(selectNF, gains)$

For this algorithm, the conditional entropy with respect to the class is computed with

$$H(X|Y) = - \sum_j P(y_j) \sum_i P(x_i|y_j) \log_2(P(x_i|y_j)), \qquad (4.4)$$

then, the information gain (IG) is computed for each attribute with

$$IG(X|Y) = H(X) - H(X|Y). \qquad (4.5)$$

Once the algorithm has all information gain values for each attribute, the top N are selected as best features.

Algorithm 5 shows the pseudocode for information gain. First the frequencies of each value with respect to the class label are computed. With this information, the total entropy of the dataset is calculated. Next, for each attribute, its frequency, probability, entropy, and conditional entropy are computed. Finally, the information gain for the i-th attribute its computed and stored in *gains*. Algorithm 6 shows the process of the frequencies calculation.

Algorithm 6 Frequencies function

Input: *attr* attribute to compute frequencies to
Input: *f* function to group by
Output: Frequencies for *attr* using *f*
 1: *grouped ← groupBy(data, f)*
 2: *freqs ← reduceGroup(grouped)*
 3: **return** *freqs*

4.5.2   OFS: Online Feature Selection

OFS [43] is an ϵ-greedy online FS method based on weights, generated by an online classifier (like neural networks) which makes a balance between exploration and exploitation of features.

The main idea behind this algorithm is that when a vector **x** falls within a $L1$ ball, most of its numerical values are concentrated in its largest elements; therefore, removing the smallest values will result in a small change in the original vector x as measured by the L_q norm. This way, the classifier is restricted to a $L1$ ball:

$$\Delta_R = \left\{ \mathbf{w} \in \mathrm{R}^d : ||\mathbf{w}||_1 \leq \mathrm{R} \right\}. \tag{4.6}$$

OFS maintains an online classifier $\mathbf{w}_t$ with at most B nonzero elements. When an instance $(\mathbf{x}_t, y_t)$ is incorrectly classified, the classifier gets updated through online gradient descent and then it is projected to a $L2$ ball to delimit the classifier norm. If the resulting classifier $\hat{\mathbf{w}}_{t+1}$ has more than B nonzero elements, the elements with the largest absolute value will be kept in $\hat{\mathbf{w}}_{t+1}$.

The above approach presents an inefficiency, even although the classifier consists of B nonzero elements, full knowledge of the instances is required, that is, each attribute $\mathbf{x}_t$ must be measured and computed. As a solution, OFS limits online FS to no more than B attributes of $\mathbf{x}_t$.

Algorithm 7 shows the pseudocode for OFS. This algorithm maps each label and feature with their corresponding value for the original OFS algorithm. Finally, it returns the selected features.

4.5.3   FCBF: Fast Correlation-Based Filter

FCBF [45] is a FS method that considers the class relevance and the dependency between each feature pair. It is based on information theory, and employs Symmetrical Uncertainty (SU) to calculate the dependency among the attributes and the class relevance. FCBF starts with the complete set of features and, heuristically, applies a backward selection method with a sequential search to remove redundant and irrelevant attributes. The algorithm stops when there are no attributes left to eliminate.

Algorithm 7 OFS algorithm

Input: *data* a DataSet LabeledVector (label, features)
Input: η parameter
Input: λ parameter
Input: *selectNF* Number of features to select
Output: DataSet with the most *selectNF* important features
1: $fs \leftarrow OFS(\eta, \lambda)$
2: *weights* $\leftarrow$
3: **map** $(label, features) \in data$
4:     $fs(label, features)$
5: **end map**
6: **return** $selectFeatures(selectNF, weights)$

The algorithm chooses as a correlation measure the entropy of a variable X, which is defined as

$$H(X) = -\sum_i P(x_i) \log P(x_i) \tag{4.7}$$

and the entropy of X after observing values of another variable Y is defined as

$$H(X|Y) = -\sum_j P(y_j) \sum_i P(x_i|y_j) \log_2(P(x_i|y_j)), \tag{4.8}$$

where $P(x_i)$ is the prior probability for all values of X and $P(x_i|y_j)$ is the posterior probability of X given the values of Y. According to IG, a feature Y is more correlated to X than to a feature Z if $IG(X|Y) > IG(Z|Y)$.

Now we are ready to define the main measure for FCBF, *Symmetrical Uncertainty* [34]. As a pre-requisite, data must be normalized in order to be comparable.

$$SU(X, Y) = 2\left[\frac{IG(X|Y)}{H(X) + H(Y)}\right] \tag{4.9}$$

SU compensates the bias in *IG* toward features with more values and normalizes its values to the range [0, 1]. A *SU* value of 1 indicates total correlation, whereas a value of 0 indicates independence.

The algorithm follows a two-step approach, first, it has to decide if a feature is *relevant* to the class and two, decide if those features are *redundant* with respect to each other.

To solve the first step, a user-defined *SU* threshold can be defined. If $SU_{i,c}$ is the *SU* value for feature F_i with the class c, the subset S' of relevant features can be defined with a threshold δ such that $\forall F_i \in S', 1 \leq i \leq N, SU_{i,c} \geq \delta$.

For the second step, in order to avoid analysis of pairwise correlations between all features, a method to decide whether the level of correlation between two features in S' is high enough to produce redundancy is needed in order to remove one of them. Examining the value $SU_{j,i} \forall F_j \in S'(j \neq i)$ allows the level to which F_j is correlated by the rest of features in S' to be estimated.

The last piece of the algorithm comprises two definitions:

Definition 4.1 (Predominant Correlation) The correlation between a feature F_i and the class C is predominant *iff* $SU_{i,c} \geq \delta$ and $\forall F_j \in S'(j \neq i) \nexists F_j$ such that $SU_{j,i} \geq SU_{i,c}$

If such feature F_j exists for a feature F_i, it is called a redundant peer of F_i and it is added to a set S_{P_i} identifying all the redundant peers for F_i. S_{P_i} is divided into two parts: $S_{P_i}^+$ and $S_{P_i}^-$, where $S_{P_i}^+ = \{F_j | F_j \in S_{P_i}, SU_{j,c} > SU_{i,c}\}$ and $S_{P_i}^- = \{F_j | F_j \in S_{P_i}, SU_{j,c} \leq SU_{i,c}\}$

Definition 4.2 (Predominant Feature) A feature is predominant to the class *if* its correlation to the class is predominant or can become predominant after removing all its redundant peers.

According to the above definitions, a feature will be a *good feature* if it is *predominant* in predicting the class. These two definitions along with the following heuristics can effectively identify predominant features and remove the need of pairwise comparisons.

Heuristic 1 (When $S_{P_i}^+ = \emptyset$) F_i is a predominant feature, delete all features in $S_{P_i}^-$ and stop searching for redundant peers for those features.

Heuristic 2 (When $S_{P_i}^+ \neq \emptyset$) All features in $S_{P_i}^+$ are processed before making decisions in F_i. If none of them become predominant go to Heuristic 1, or else remove F_i and decide if features in $S_{P_i}^-$ need to be removed based on other features in S'.

Heuristic 3 (Start Point) The algorithm begins examining the feature with the largest $SU_{i,c}$, as this feature is always predominant and acts as a starting point for the removal of redundant features.

Algorithm 8 shows pseudocode for FCBF, the SU value is computed for each attribute in parallel. All SU values are then filtered according to the threshold parameter and then sorted in descending order. With these final sorted values, the FCBF algorithm is applied as originally described in [45].

Algorithm 9 shows how Symmetrical Uncertainty is computed in a distributed fashion. First, each parallel partition computes the partial counts of each value, then this partial counts are aggregated using a reduce function in order to compute the total counts. With this information, probabilities for each value are computed and the entropy and mutual information are calculated. Finally, it returns the corresponding SU value for that attribute.

In classic scenarios, when performing dimensionality reduction the full set of data is available. Big Data streaming is a special task, as the full data is not available, and the available data can be seen only for a short amount of time. For this reason, classic data reduction algorithms are not able to tackle this problem. In this section we have shown a library, composed of three data reduction algorithms focused on

---

Algorithm 8 FCBF algorithm

Input: *data* a DataSet LabeledVector (label, features)
Input: *thr* threshold
Output: DataSet with the most important features
 1: $su \leftarrow$
 2: **for** $i = 0$ *until nAttrs* **do**
 3:     $attr \leftarrow$
 4:     **map** *instance* $\in$ *data*
 5:        $(label, feature_i)$
 6:     **end map**
 7:     **yield** $SU(attr)$
 8: **end for**
 9: $suSorted \leftarrow filter(su > thr).SortDesc$
10: $sBest \leftarrow FCBF(suSorted)$
11: **return** $sBest$

Algorithm 9 Symmetrical Uncertainty (SU) function

Input: *attr* Attribute to compute SU to
Output: SU value for *attr*
 1: $xypartialCounts \leftarrow$
 2: **map partitions** $(y, x) \in attr$
 3:     $xPartialCounts \leftarrow computeCounts(x)$
 4:     $yPartialCounts \leftarrow computeCounts(y)$
 5:        $(xPartialCounts, yPartialCounts)$
 6: **end map**
 7: $totalCounts \leftarrow reduce(xypartialCounts)$
 8: $su \leftarrow$
 9: **map** $(xcounts, ycounts, x, y) \in totalCounts$
10:     $px \leftarrow prob(x)$
11:     $py \leftarrow prob(y)$
12:     $hx \leftarrow entropy(x)$
13:     $hy \leftarrow entropy(y)$
14:     $mu \leftarrow mutualInformation(x, y)$
15:     $\frac{2mu}{hx+hy}$
16: **end map**

Big Data streams named DPASF. This library uses Apache Flink as its Big Data streaming core framework. The proposed algorithms are able to evolve with the data, and to select the best features.

4.6 Summary and Conclusions

Dimensionality reduction is the set of techniques devoted to reduce the number of features in the data. More features are not always better, as they may contain noise or be highly correlated with others. Storage reduction is also an important side of dimensionality reduction.

In this chapter we have reviewed the most recent techniques for dimensionality reduction for both batch and streaming data. For static data, one of the most powerful solutions is a framework with several information theory-based FS methods. On the other hand, for streaming data, we have analyzed three methods that perform FS in Big Data streaming scenarios.

The Big Data explosion presents some challenges for the dimensionality reduction task, but it also brings new opportunities [24]. Big Data problems with ultra-high dimensionality are becoming more common. This new challenge can be found in many real-world applications, such as text mining or information retrieval. Big Data dimensionality reduction algorithms may not scale well enough to tackle this amount of features. For these problems, new algorithms with linear or sublinear running times will be required.

Regarding dimensionality reduction in Big Data streaming environments, little research has been devoted to it so far. There are still many open challenges, like the selection of relevant and timely features in just one pass over the data, which will require the future efforts of experts in the field. While Big Data streaming is becoming more and more popular, the necessity for dimensionality reduction algorithms for selecting the best features in the data stream is more and more pressing everyday.

References

1. Aggarwal, C. C. (2015). *Data mining: the textbook*. Berlin: Springer.
2. Alcalde-Barros, A., García-Gil, D., García, S., & Herrera, F. (2019). DPASF: A Flink library for streaming data preprocessing. *Big Data Analytics, 4*(1), 4.
3. Apache Flink. (2019). *Apache Flink*. http://flink.apache.org/.
4. Battiti, R. (1994). Using mutual information for selecting features in supervised neural net learning. *IEEE Transactions on Neural Networks, 5*(4), 537–550.
5. Bolón-Canedo, V., Sánchez-Maroño, N., & Alonso-Betanzos, A. (2015). Recent advances and emerging challenges of feature selection in the context of big data. *Knowledge-Based Systems, 86*, 33–45.
6. Bolón-Canedo, V., Sánchez-Maroño, N., & Alonso-Betanzos, A. (2015). *Feature selection for high-dimensional data*. Berlin: Springer Publishing Company, Incorporated.
7. Bondell, H. D. & Reich, B. J. (2008). Simultaneous regression shrinkage, variable selection, and supervised clustering of predictors with OSCAR. *Biometrics, 64*, 115–123.
8. Brown, G., Pocock, A., Zhao, M.-J., & Luján, M. (2012). Conditional likelihood maximisation: A unifying framework for information theoretic feature selection. *Journal of Machine Learning Research, 13*, 27–66.
9. Chang, C.-C., & Lin, C.-J. (2011). LIBSVM: A library for support vector machines. *ACM Transactions on Intelligent Systems and Technology, 2*, 27:1–27:27. Software available at http://www.csie.ntu.edu.tw/~cjlin/libsvm
10. Chao, P., Bin, W., & Chao, D. (2012). Design and implementation of parallel term contribution algorithm based on MapReduce model. In *7th Open Cirrus Summit* (pp. 43–47). Piscataway: IEEE.
11. Chen, K., Wan, W. Q., & Li, Y. (2013). Differentially private feature selection under MapReduce framework. *The Journal of China Universities of Posts and Telecommunications, 20*(5), 85–103.

12. Cover, T. M., & Thomas, J. A. (1991). *Elements of information theory*. Hoboken: Wiley-Interscience
13. Dalavi, M., & Cheke, S. (2014). Hadoop MapReduce implementation of a novel scheme for term weighting in text categorization. In *International Conference on Control, Instrumentation, Communication and Computational Technologies (ICCICCT)* (pp. 994–999). Piscataway: IEEE.
14. del Río, S., López, V., Benítez, J. M., & Herrera, F. (2014). On the use of MapReduce for imbalanced big data using random forest. *Information Sciences, 285*, 112–137.
15. Díaz-Uriarte, R., & de Andrés, S. A. (2006). Gene selection and classification of microarray data using random forest. *BMC Bioinformatics, 7*(1), 3.
16. Dua, D., & Graff, C. (2017). *UCI machine learning repository*. http://archive.ics.uci.edu/ml.
17. Fung, G. M., & Mangasarian, O. L. (2004). A feature selection newton method for support vector machine classification. *Computational Optimization and Applications, 28*(2), 185–202.
18. Gehler, P., & Nowozin, S. (2009). On feature combination for multiclass object classification. In *2009 IEEE 12th International Conference on Computer Vision* (pp. 221–228).
19. Guyon, I., Weston, J., Barnhill, S., & Vapnik, V. (2002). Gene selection for cancer classification using support vector machines. *Machine Learning, 46*(1), 389–422.
20. He, Q., Cheng, X., Zhuang, F., & Shi, Z. (2014). Parallel feature selection using positive approximation based on MapReduce. In *11th International Conference on Fuzzy Systems and Knowledge Discovery FSKD* (pp. 397–402).
21. Hodge, V. J., OKeefe, S., & Austin, J. (2016). Hadoop neural network for parallel and distributed feature selection. *Neural Networks, 78*, 24–35. In press, https://doi.org/10.1016/j.neunet.2015.08.011.
22. Katakis, I., Tsoumakas, G., & Vlahavas, I. (2005). On the utility of incremental feature selection for the classification of textual data streams. In P. Bozanis, & E. N. Houstis (Eds.), *Advances in informatics* (pp. 338–348). Berlin: Springer.
23. Kumar, M., & Rath, S. K. (2015). Classification of microarray using MapReduce based proximal support vector machine classifier. *Knowledge-Based Systems, 89*, 584–602.
24. Li, J., & Liu, H. (2017). Challenges of feature selection for big data analytics. *IEEE Intelligent Systems, 32*(2), 9–15.
25. Li, Z., Lu, W., Sun, Z., & Xing, W. (2017). A parallel feature selection method study for text classification. *Neural Computing and Applications, 28*(1), 513–524.
26. Mao, Q., & Tsang, I. W. (2013). Efficient multitemplate learning for structured prediction. *IEEE Transactions on Neural Networks and Learning Systems, 24*, 248–261.
27. Meena, M. J., Chandran, K. R., Karthik, A., & Samuel, A. V. (2012). An enhanced ACO algorithm to select features for text categorization and its parallelization. *Expert Systems with Applications, 39*(5), 5861–5871.
28. Michael, M., & Lin, W.-C. (1973). Experimental study of information measure and inter-intra class distance ratios on feature selection and orderings. *IEEE Transactions on Systems, Man and Cybernetics, SMC-3*(2), 172–181.
29. Nie, F., Xiang, S., Jia, Y., Zhang, C., & Yan, S. (2008). Trace ratio criterion for feature selection. In *Proceedings of the 23rd National Conference on Artificial Intelligence - Volume 2*, AAAI'08 (pp. 671–676).
30. O'Leary, D. E. (2013). Artificial intelligence and big data. *IEEE Intelligent Systems, 28*, 96–99 (2013)
31. Ordozgoiti, B., Gómez-Canaval, S., & Mozo, A. (2015). Massively parallel unsupervised feature selection on spark. In *New trends in databases and information systems*. Communications in Computer and Information Science (Vol. 539, pp. 186–196). Berlin: Springer International Publishing.
32. Peng, H., Long, F., & Ding, C. (2005). Feature selection based on mutual information criteria of max-dependency, max-relevance, and min-redundancy. *IEEE Transactions on Pattern Analysis and Machine Intelligence, 27*(8), 1226–1238.

33. Peralta, D., del Río, S., Ramírez, S., Triguero, I., Benítez, J. M., & Herrera, F. (2015). Evolutionary feature selection for big data classification: A MapReduce approach. *Mathematical Problems in Engineering, 2015*, Article ID 246139.
34. Press, W. H., Flannery, B. P., Teukolsky, S. A., & Vetterling, W. T. (1988). *Numerical recipes in C: The art of scientific computing.* New York: Cambridge University Press.
35. Ramírez-Gallego, S., Mouriño-Talín, H., Martínez-Rego, D., Bolón-Canedo, V., Benítez, J. M., Alonso-Betanzos, A., et al. (2018). An information theory-based feature selection framework for big data under apache spark. *IEEE Transactions on Systems, Man, and Cybernetics: Systems, 48*(9), 1441–1453.
36. Roush, W. (2019). *MIT technology review.* TR10: Peering into video's future. http://www.technologyreview.com/Infotech/18284/
37. Singh, S., Kubica, J., Larsen, S. E., & Sorokina, D. (2009). Parallel large scale feature selection for logistic regression. In *SIAM International Conference on Data Mining (SDM)* (pp. 1172–1183).
38. Spark, A. (2019). *Machine learning library (MLlib) for spark.* http://spark.apache.org/docs/latest/mllib-guide.html.
39. Sun, Z., & Li, Z. (2014). Data intensive parallel feature selection method study. In *International Joint Conference on Neural Networks (IJCNN)* (pp. 2256–2262).
40. Tan, M., Tsang, I. W., & Wang, L. (2014). Towards ultrahigh dimensional feature selection for big data. *Journal of Machine Learning Research, 15*(1), 1371–1429.
41. Tanupabrungsun, S., & Achalakul, T. (2013). Feature reduction for anomaly detection in manufacturing with MapReduce GA/kNN. In *19th IEEE International Conference on Parallel and Distributed Systems (ICPADS)* (pp. 639–644).
42. Triguero, I., del Río, S., López, V., Bacardit, J., Benítez, J. M., & Herrera, F. (2015). ROSEFW-RF: The winner algorithm for the ECBDL'14 big data competition: An extremely imbalanced big data bioinformatics problem. *Knowledge-Based Systems, 87*, 69–79.
43. Wang, J., Zhao, P., Hoi, S. C. H., & Jin, R. (2014). Online feature selection and its applications. *IEEE Transactions on Knowledge and Data Engineering, 26*(3), 698–710.
44. Yazidi, J., Bouaguel, W., & Essoussi, N. (2016). *A parallel implementation of relief algorithm using MapReduce paradigm* (pp. 418–425). Cham: Springer International Publishing.
45. Yu, L., & Liu, H. (2003). Feature selection for high-dimensional data: A fast correlation-based filter solution. In *Proceedings of the 20th International Conference on Machine Learning (ICML-03)* (pp. 856–863).
46. Zdravevski, E., Lameski, P., Kulakov, A., Jakimovski, B., Filiposka, S., & Trajanov, D. (2015). Feature ranking based on information gain for large classification problems with MapReduce. In *2015 IEEE Trustcom/BigDataSE/ISPA* (Vol. 2, pp. 186–191).
47. Zhai, Y., Ong, Y.-S., & Tsang, I. W. (2014). The emerging "big dimensionality". *IEEE Computational Intelligence Magazine, 9*(3), 14–26.
48. Zhai, Y., Tan, M., Ong, Y. S., & Tsang, I. W. (2012). Discovering support and affiliated features from very high dimensions. In J. Langford & J. Pineau (Eds.), *Proceedings of the 29th International Conference on Machine Learning (ICML-12)* (pp. 1455–1462). New York: ACM.
49. Zhao, Z., Zhang, R., Cox, J., Duling, D., & Sarle, W. (2013). Massively parallel feature selection: an approach based on variance preservation. *Machine Learning, 92*(1), 195–220.
50. Zhong, L. W., & Kwok, J. T. (2012). Efficient sparse modeling with automatic feature grouping. *IEEE Transactions on Neural Networks and Learning Systems, 23*(9), 1436–1447.

Chapter 5
Data Reduction for Big Data

5.1 Introduction

Data reduction techniques [19] emerged as preprocessing algorithms that simplify and clean raw data early in the early stages while retaining as much information as possible. These techniques are used to both obtain a representative sample of the original data and alleviate data storage requirements [41]. This process does not only obtain a relevant sample of the original data, but also aims at eliminating noisy instances, and redundant or irrelevant data, improving the later data mining (DM) process.

In the literature, there are two main approaches to perform data reduction, consisting of reducing the number of input attributes or the instances. Focusing on reducing attributes, the most popular data reduction techniques are FS and feature extraction [28], which are designed to either select the most representative features or construct a new whole set of them. Similarly, from the instances point of view, we can differentiate between instance selection (IS) methods [18] and instance generation (IG) methods [38]. The objective of an IS method is to obtain a subset $SS \subset TR$ such that SS does not contain redundant or noisy examples and $Acc(SS) \simeq Acc(TR)$, where $Acc(SS)$ is the classification accuracy when using SS as the training set. Likewise, IG methods may generate artificial data points if needed for a better representation of the training set. The purpose of an IG method is to obtain a generated set IGS, which consists of p, $p < n$, instances, which can be either selected or generated from the examples of TR.

Most existing instance reduction methods were actually conceived to address the weaknesses of the k-nearest neighbors (KNN) algorithm [10]. These methods are known as prototype reduction (PR) techniques. Prototype selection (PS) methods are IS methods that use an instance-based classifier with a distance measure, commonly KNN, for finding a representing subset of the training set. Classical selection methods, such as reduced-NN, edited-NN, or condensed-NN, utilize the NN rule to evaluate instances. Here the neighbor set helps to decide if a given instance is relevant (must be preserved), or redundant or noisy (to be removed). One of the classic and most widely used algorithms for PS is the fast condensed nearest neighbor (FCNN), which is an order-independent algorithm to find a consistent subset of the training dataset using the NN rule [1]. Another simple yet powerful example is the random mutation hill climbing (RMHC) [37]; it randomly selects a subset of the training data and performs RMHC iteratively to select the best subset using KNN as a classifier. The IS problem can be seen as a binary optimization problem which consists of whether or not to select a training example [15]. For this reason, evolutionary algorithms have been used for PS, with very promising results. In these algorithms, the fitness function usually consists of classifying the whole training set using the KNN algorithm [5]. To date, one of the best performing algorithms for evolutionary PS is [17], which is a steady-state memetic algorithm (SSMA) that achieves a good reduction rate and accuracy with respect to classical PS schemes.

Another approach to perform instance reduction is IG, also called prototype generation (PG) in the case of instance-based classifiers. In contradistinction to PS, these methods aim to overcome an additional limitation of the KNN algorithm: it makes predictions over existing training data assuming they perfectly delimit the decision boundaries between classes (in classification problems). To overcome that limitation, these methods are not restricted to selecting examples of the training data, but they can also modify the values of the instances based on nearest neighbors. The most popular strategy is to use merging of nearest examples to set the new artificial samples [8]. We can also find clustering-based approaches [4] or evolutionary-based schemes [39], but the vast majority of them are based on the idea of computing nearest neighbors to reduce the training set. A complete survey on this topic can be found in [38].

In terms of FS, a variety of strategies such as wrappers, filters, and embedded methods have been proposed in the literature [25]. Nevertheless, we can still find that the KNN algorithm has also played an important role in many existing FS proposals [31]. One of the classic and most relevant methods is ReliefF [26] that ranks features according to how well an attribute allows us to distinguish the nearest neighbors within the same class label from the nearest neighbors from each of the different class labels. Similarly to the instance reduction scenario, evolutionary algorithms have also been employed to perform FS with good results. In [45] we can find a complete survey on FS using evolutionary computation.

Hybrid approaches for data reduction have also been proposed in the literature. Instead of using IS and FS methods separately, some research has been devoted to the combination of both IS and FS. In [12], for instance, a hybrid of IS and FS algorithm is presented, using an evolutionary model to perform FS and IS for KNN classification. Hybrid approaches of PS and PG have also been studied in the literature. In these methods, PS is used for selecting the most representative subset of the training data, and PG is tasked to improve this subset by modifying the values of the instances. In [40] a hybrid combination of SSMA with a scale factor local search in differential evolution (SSMA-SFLSDE) is introduced.

As such, PR techniques should ease the later DM processes or actually making them possible in the case of Big Data problems. However, these methods are also affected by the increase of the size and complexity of data, being unable to provide a preprocessed dataset in a reasonable time. Several solutions have been developed to enable data reduction techniques to deal with this problem.

We can find a data level approach (called stratification) that is based on a parallel partitioning model with equal distribution of classes in partitions. This splits the original training data into several subsets that are independently addressed. Afterwards, all partial reduced subsets are concatenated to form a final solution. This approach has been used for IS [6, 13] and IG [43] with outstanding results.

However, when stratification faces large-scale problems, several major complexity matters arise. For instance, the partitioning and reduction steps become prohibitive since either they require too much memory or computing time. Furthermore, stratification does not consider redundancy/noise relationships arising after the aggregation phase.

On the other hand, in many cases we do not only deal with static data collections, but rather with dynamic ones. They arrive in a form of continuous batches of data, called data streams [16]. In this dynamic scenario, we need not only to manage the volume, but also the velocity of data, constantly updating the model to the current state of the stream. To add a further difficulty, many modern data sources generate their outputs with very short intervals, thus leading to high-speed data streams [23]. To enable the application of lazy learning to streaming environments, we need PR methods that prevent the accumulation of instances and that outdated concepts are utilized to make erroneous decisions.

In the remaining parts, we shall begin with the first developments on scalability for sequential PR (Sect. 5.2). We shall continue with the analysis of one of the few proposals in the literature that address the problem of large-scale PR (Sect. 5.3). Afterwards, we analyze the proposal of a library with several methods for PR for Big Data scenarios (Sect. 5.4). Then, we shall study how PR can be utilized to ease the ingestion of data in high-speed streaming systems (Sect. 5.5). Finally, Sect. 5.6 summarizes the chapter and gives some conclusions.

5.2 Parallel and Rapid Prototype Reduction

As mentioned before, PR arises as a great tool to tackle large set of instances. However, in practice most of methods are unable to be applied, or even if they could be executed they would only generate wrong results [21].

To solve this scalability problem, several parallel models have been proposed to improve the scalability of PR methods. The first contribution to this end has been a data stratification technique which splits the original data into a range of partition with equal distribution of classes [34]. This strategy was then implemented for evolutionary PR [6, 7] and memetic PR [13], among others.

Another alternative proposed in the literature has been the divide-and-conquer model, which perfectly adjusts to distributed processing models like MapReduce. García-Osorio et al. [20] were pioneer in this field. In their model, the initial training set is divided into a set of disjoint subsets distinct in each iteration. An individual PS process is applied on each subset, and each removable instance receives a vote. After the upper limit of iterations is exceeded, instances with the largest number of votes are eliminated. A similar approach was proposed in [22], in which each subset is reduced using a PS technique and the partial results are combined (by voting) into a final set. A more complex divide-and-conquer policy is introduced in [11]. This recursive algorithm divides the training set, applies a PR algorithm, and then, aggregates the partial results into a subset which will serve as a starting point for the next iteration. It also introduces a cross validation process that allows the learning model to be improved until the validation error grows.

As introduced in [20], the split procedure is highly dependent on the problem addressed. Splitting initial set into small subsets for its posterior processing implies a subsequent decrease in accuracy due to a more restricted vision of the original problem. While it could be addressed by forcing equality in class distributions in every subset, this becomes unfeasible in large-scale scenarios where overall data distribution is theoretically unknown.

Besides the aforementioned problem, there exist some extra issues that appear when we increase the number of instances, which are mainly three-folded:

- Stratification does not consider that joining each partial solution into a global one could generate a reduced set with redundant or noisy instances, which is the normal scenario. This nuance may damage the classification performance of the subsequent learning process.
- A stratified partitioning process could not be carried out when the dataset does not completely fit in main memory.
- The complexity associated with PR is mainly quadratic or greater, which is excessive for real-world big applications.

A sequential approach has been proposed in [2] that solve most of problems mentioned before. This PS technique (called LSH-IS) utilizes locality-sensitive hashing to assign similar prototype to the same bucket in linear time. Additionally, LSH-IS does not require the entire set to be in memory, but this can be run incrementally. While LSH-IS represents a quantum leap on performance for standard PR, this strategy cannot be directly embraced by distributed environments.

In [33], authors propose an approach for FS based on KNN for Big Data, where FS is performed on huge datasets using the KNN algorithm within an evolutionary approach. More recently, a distributed Spark-based version of the ReliefF algorithm has been presented [32].

In [3], Arnaiz-González et al. present a parallelized MapReduce version of the democratic instance selection algorithm, called MR-DIS. This method partitions the original data into several disjoint subsets and applies an independent PS process on each subset. This process is repeated several rounds in which each removed instance receives a single vote. The subsequent counters are utilized to decide which voting threshold is the most appropriate for the target problem according to a predefined fitness function. While [43] can be deemed as a fully operating solution, able to scale properly in distributed environments, it does not provide further improvement with respect to the original algorithm.

5.3 MRPR: A MapReduce Solution for Prototype Reduction in Big Data Classification

In this section, we will describe the MRPR framework (MapReduce for prototype reduction), a new distributed framework for PR based on the stratification procedure [43]. This framework was designed to tackle the drawbacks associated with stratification: high memory consumption and high complexity, and a poor joining process. MRPR relies on MapReduce to parallelize the PR process and the subsequent fusion process. Concretely, the map phase contains the splitting procedure and the local application of PR. The reduce stage performs a filtering or fusion of prototypes in order to prevent the inclusion of negative prototypes in the final set. Figure 5.1 depicts a simplified scheme of the MRPR framework.

Fig. 5.1 MRPR processing scheme. The rectangles represent the reduction and joining processes, and the circles the partial and final reduced sets

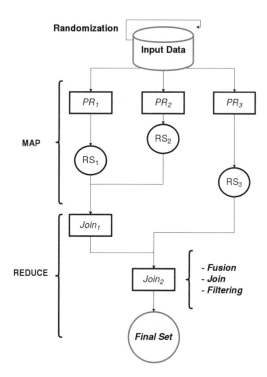

Although MRPR utilizes several maps to achieve a full parallelization of reduction, it only uses a long-live reduce process that is updated every time a mapper is completed. With the adopted strategy, a single reduce in this scheme has shown to be less time-consuming than multiple simultaneous reducers. This improvement has shown a decrease in the overall MapReduce overhead, especially in network usage [9].

Note that the main improvement introduced by MRPR with respect to strati-fication resides in the more sophisticated aggregation procedure implemented in the reduce stage. This procedure aims at reducing the inconsistencies derived from partial views. In counterpart to MRPR, partial outcomes are simply concatenated in stratification.

In the following, we describe the single MapReduce phase implemented in MRPR. We start describing the map function where the PR process occurs, and then the joining process performed in the reduce phase is studied.

5.3.1   Map Stage: Distributed Prototype Reduction

The first step of MRPR is devoted to read the training set (TR) from a distributed file system, like Hadoop. This set is formed by several disjoint subsets of instances (TR_i) which correspond with the HDFS blocks read (following the Hadoop example). Data partitions are then evenly distributed across the m map tasks so that each map will process approximately the same number of instances.

Under this scheme, if the partitioning procedure is directly applied over TR, the class distribution of each partition could be biased to the original distribution of instances. Additionally, a proper stratified partitioning could not be performed as stated previously. In order to guarantee that partitions share the same proportion of classes, we randomly shuffle the instances in TR across the cluster. This operation is quite lightweight in comparison with the application of the PR technique, and it will be applied once. Although non-deterministic, randomization guarantees the proportion of classes is approximately maintained in all partitions.

Once each map has formed its corresponding training set, a PR step is performed on this set as a whole. This function basically defines the application of the PR technique on each training partition. As a result, each map generates a partial subset (RS_i) as output.

Note that PR techniques may perform differently depending on the main characteristics of the dataset, and the stopping criteria held by each PR algorithm. In general, map tasks will end closely because differences between partitions are marginal due to randomization. In any case, MapReduce starts the reduce phase when the first mapper ends. As each map finishes its processing the results are forwarded to the single reduce task.

5.3.2   Reduce Stage: Aggregation of Partial Results

The reduce phase aims at aggregating all the reduce subsets generated into a single set. This process is iterative and can be implemented following different fusion models:

- **Join**: This baseline option concatenates all the reduced sets into a final set. Although simple, this joining process does not guarantee that the resulting set does not contain irrelevant or even noisy instances. In this case, the intervention performed by the reduce stage is minimum. However, this model will serve as a starting point for further developments.
- **Filtering**: This alternative exploits the idea of filtering noisy instances during the aggregation. This filtering technique is based on the edition family of methods [19], which use simple heuristics to discard points that are noisy or do not agree with their neighbors. They provide smoother decision boundaries for KNN. In general, edition schemes improve generalization by performing a slight reduction on TR. This behavior fits perfectly the objective of the reduce phase in MRPR, since in this stage we do not pursue to reduce more, but only removing noisy instances.

  The reduce function iteratively applies a filtering on the current set. It means that as the mappers end their execution, the reduce function starts to gradually aggregate subsets by applying filtering until all the partial results are collected. Finally, a single filtered set is obtained.
- **Fusion**: This variant gears towards eliminating redundant prototypes. In this case, fusion is based on the centroid-based family of methods for PR. These techniques merge similar examples to obtain a less populated set of instances. Given that in this step we want to generate a single subset without redundancies, these methods can be very useful. As in the previous scheme, the fusion stage will be progressively applied during the creation of RS.

5.3.3   On the Election of Prototype Reduction Methods for Distributed Systems

In this part, we analyze some directives to select the most suitable PR technique for each target problem. We also discuss which PR techniques are more convenient for the map and reduce stages, and what are the relationships between them.

As we mentioned before, PR is applied locally in the map phase to select a subset of relevant instances. Then, depending on the PR model used in mappers we should select a filtering or a fusion PR technique to combine the resulting reduced sets. Although all PR methods in the literature could fit in the map phase, we must consider some aspects (reduction, accuracy, and runtime) before proceeding with selection. Firstly, a very accurate PR technique is desirable. However, this kind of techniques usually implies a low reduction rate, and therefore, a negative impact in the performance of the reduce phase (join). Secondly, depending on the complexity of the PR method, we should vary the number of mappers to divide the original set. The more complex the method, the more mappers are required. Note that the latter fact implies a considerable reduction on the representativeness of subsets, and a subsequent drop in accuracy.

According to [19], there exist six main families of PR methods: edition, condensation, hybrid approaches, positioning adjustment, centroids-based, and space splitting. Each one has associated a list of recommendations to be considered before designing a MRPR job:

- Edition-based methods are focused on cleaning input data by removing noisy instances. Thus, these methods are normally fast and accurate but they obtain a quite low reduction rate. To implement these methods in MRPR, it is recommended a rapid joining phase in reducers.
- Condensation, hybrid, and space splitting algorithms typically offer a fair trade-off between reduction, accuracy, and runtime. Their reduction rate in these methods is quite fluctuating, so depending on the problem faced, the reducer chosen should be more or less efficient.
- Positioning adjustment are highly customizable techniques that may be used to apply a deep reduction on input data. These techniques can provide very accurate results in a relatively moderate runtime. The high reduction rate held by them will allow us to apply an accurate reduction in the reduce phase without a meaningful loss in performance.
- Centroid-based algorithms are accurate but in general quite time-consuming. Although its integration is feasible and could be useful in some scenarios, their utilization is only recommended for the reduce phase.

Although MRPR framework can perform a wide variety of PR methods in Big Data environments, we will focus on the hybrid SSMA-SFLSDE algorithm [40] to test the MRPR model. Furthermore, the ENN algorithm [44] is used as edition method for the filtering-based reducer. For the fusion-based reducer, a very accurate centroid-based technique called ICLP2 [27] is applied. It is motivated by the high reduction ratio of these positioning adjustment methods. In addition, the NN classifier has been included as baseline limit of performance.

The dataset employed in the test is SUSY dataset [14]. This dataset is composed of 5,000,000 instances, 18 attributes, and 2 classes. It has been partitioned using a fivefold cross validation scheme. It means that the dataset is split into five folds, each one containing 20% of the examples of the dataset. For each fold, a PR algorithm is run over the examples presented in the remaining folds (that is, in the training partition, TR). Then, the resulting RS is tested with the current fold using the NN rule. Test partitions are kept aside during the PR phase in order to analyze the generalization capabilities provided by the generated RS. Because of the randomness of some operations that these algorithms perform, they have been run three times per partition.

Table 5.1 Results obtained for the SUSY problem

Reduce type	# Maps	Training	Test	Runtime	Reduction rate	Classification time
Join	256	**69.53**	**72.34**	69,153	97.00	30,347
Filtering	256	**69.41**	**72.82**	66,371	97.77	24,686
Fusion	256	68.70	**72.40**	69,797	98.91	11,422
Join	512	68.96	**72.17**	26,011	97.21	35,068
Filtering	512	68.98	**72.41**	28,508	97.56	24,868
Fusion	512	68.10	**72.30**	30,344	98.83	12,169
Join	1024	**69.39**	**71.88**	13,525	97.15	45,388
Filtering	1024	68.26	**72.26**	14,511	97.32	32,568
Fusion	1024	67.57	**72.08**	15,562	98.70	12,136
NN	–	68.99	71.57	–	–	1,167,200

Table 5.1 summarizes all the results obtained on SUSY dataset. It shows training/test accuracy, runtime, and reduction rate obtained by the SSMA-SFLSDE algorithm, in the MRPR framework, depending on the number of mappers (# Maps) and reduce type. For each one of these measures, average results are presented (from the 5-fcv experiment). Moreover, the average classification time in the TS is computed as the time needed to classify all the instances of TS with the corresponding RS generated by MRPR. Furthermore, we compare these results with the accuracy and test classification time achieved by the NN classifier. It uses the whole TR set to classify all the instances of TS. Average accuracies higher or equal than the obtained with the NN algorithm have been highlighted in bold.

The experimental study carried out has shown that the MRPR framework obtains very competitive results. It allows to apply PR techniques in large-scale problems. It is able to improve the accuracy with a reduction of 98% less instances. The application of the MRPR framework has resulted in a very big reduction of storage requirements and classification time for the NN rule when dealing with big datasets.

5.4   Transforming Big Data into Smart Data Through Data Reduction

In this section, we will describe the PR methods proposed in   [41]. In this paper, authors discuss the role of one of the simplest DM techniques—the KNN algorithm—as a powerful tool to obtain "Smart Data," which is data with a high quality to be mined.

Authors analyze the behavior of some of the most representative instance reduction approaches based on KNN when tackling Big Data datasets. MR-DIS and SSMA-SFLSDE were already proposed for Big Data as local models (apply IS and IG algorithms in different chunks of data). For the experiments, the FCNN has been adapted to Big Data using the MRPR framework [43] (same framework used for

SSMA-SFLSDE). MRPR and MR-DIS follow a local approach, which means that these methods will operate on separated chunks of data. Due to its simplicity, the RMHC algorithm has been implemented in a global manner based on the KNN-IS [30], so that, it looks at the training data as a whole (although it looks at the data taking iteratively subsets of the whole dataset).

- FCNN_MR [1]: This method applies FCNN locally in separate chunks of the data, using the MRPR framework [43]. FCNN begins with the centroids of the different classes as initial subset. Then, each iteration, for every instance in the subset, it adds the nearest enemy inside its Voronoi region. This process is repeated until no more instances are added to the subset. The resulting reduced sets from each chunk are joined together.
- MR-DIS [3]: It applies a condensed nearest neighbor algorithm [24] repeatedly to each partition of the data (locally). After each round, selected instances receive a vote. The instances with the most votes are removed.
- SSMA-SFLSDE [40]: Following a local MRPR approach, this performs an IS phase to select the most representative instances per class. Then the particular of positioning of prototypes is optimized with a differential evolution algorithm. The resulting partial reduced sets are joined together.
- RMHC_MR [37]: This is implemented as a global model. It starts from a random subset, and at each iteration, a random instance from the sample is replaced by another from the rest of the data. If the classification accuracy is improved (using the global KNN), the new sample is maintained for the next iteration.

In Fig. 5.2 we can find all the implementations of the techniques analyzed in this section, available as a Spark Package for public use.

Although these PR algorithms have been tested using seven Big Data classification problems, we are focusing on the most notorious results from three of them. These datasets are extracted from the UCI machine learning and KEEL datasets repositories [14, 42]. Table 5.2 presents the number of examples, number of features,

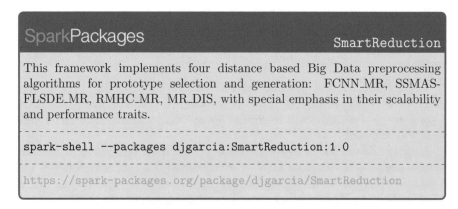

Fig. 5.2 Spark package: SmartReduction

Table 5.2 Summary description of the datasets

Dataset	# Examples	# Features	#ω
Ht-sensor	928,991	11	3
Skin	245,057	3	2
SUSY	5,000,000	18	2

Table 5.3 Impact of instance reduction on KNN (Test accuracy)

Dataset	Method				
	Baseline	FCNN_MR	SSMA-SFLSDE	MR-DIS	RMHC_MR
Ht_sensor	99.99	68.97	99.71	95.85	99.97
Skin	99.95	99.87	99.76	99.86	99.90
SUSY	69.35	66.70	70.80	67.24	67.59

and the number of classes (#ω) for each dataset. All datasets have been partitioned using a fivefold cross validation scheme. This means that each partition includes 80% of samples for training and 20% of them are left out for test.

To assess the performance of the experimental study, the accuracy and reduction rate are used. For classification, the KNN algorithm is employed. As recommended by different authors, we have focused on a $k = 1$ for the PR methods and the classification algorithm.

The cluster used for all the experiments performed in this work is composed of 14 nodes managed by a master node. All nodes have the same hardware and software configuration. Regarding the hardware, each node has 2 Intel Xeon CPU E5-2620 processors, 6 cores (12 threads) per processor, 2 GHz and 64 GB of RAM. The network used is InfiniBand 40 Gb/s. The operating system is Cent OS 6.5, with Apache Spark 2.2.0. and the maximum number of concurrent operations is equal to 256 and 2 GB for each task.

In Table 5.3, we can find the test accuracy results and reduction rate using the KNN algorithm as a classifier. Baseline represents the results of the KNN algorithm without any preprocessing. As we can see, none of the data reduction algorithms methods is losing that much accuracy with respect to the baseline accuracy. In fact, in some cases they are able to improve the baseline performance, as they remove redundant and also noisy examples. There is not a clear outperforming method overall. The choice of the right technique crucially depends on the particular problem, and the needs to reduce data storage requirements and precision. In SUSY dataset, SSMA-SFLSDE is improving the baseline accuracy by 1.5% with close to 96% of reduction. This exemplifies the importance of using data reduction techniques, not only for reducing the size of the data, but also for removing noisy and redundant instances. For datasets with high accuracy such as Skin and Ht_sensor, we can achieve up to 98.6% of reduction without losing accuracy. This allows techniques that could not be applied due to the size of the data, to be used in subsequent processes.

As a way of quantifying the reduction rate impact, Fig. 5.3 plots the data reduction rate for all tested PR methods on the SUSY dataset.

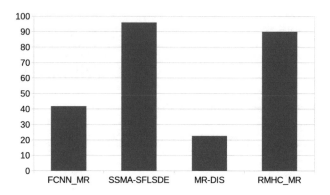

Fig. 5.3 Reduction rate (%) for SUSY dataset

The main goal of this type of techniques is to widely reduce the amount of data samples that we keep as training data. However, the analyzed algorithms work quite differently and they lead to very different accuracy and reduction rates. As we can see, for the same dataset, depending on the technique used, the reduction rate may vary from 22 to 96% of reduction. This shows the importance of choosing the right technique depending on whether the objective is to reduce data size or the focus is on obtaining a high accuracy.

5.5 Nearest Neighbor Classification for High-Speed Big Data Streams Using Spark

In this section, we will describe the algorithm DS-RNGE (data stream—relative neighborhood graph edition), a lazy learning solution for massive data streams [35]. This consists of a distributed case-base, and a PS method inspired by the relative neighborhood graph edition (RNGE) algorithm [36]. In RNGE, instances are removed if they disagree with their neighbors in a special proximity graph called relative neighborhood graph.

Since an ever-growing and noisy case-base is unacceptable in streaming environments, an improved local version of RNGE was introduced to control the insertion and removal of noisy instances. In DS-RNGE, the original RNGE was redesigned for incremental learning. In this new version, a relative graph is built around each incoming instance, and its neighbors. All the local graphs generated for new instances are then used to edit the case-base by deciding what instances should be inserted, removed, or conserved. As every step in this process is performed locally, the communication overhead is negligible.

Improving the linear search implemented in KNN, DS-RNGE utilizes a distributed metric tree to organize its case-base. Metric trees smartly index data through a metric-space explicit ordering [29]. They exploit properties such as the triangle inequality to make searches much more efficient in average, skipping a great amount of comparisons. Notice that the entire tree is maintained in memory to expedite further neighbor queries in Spark.

The distributed structure of the tree is described as follows. A single top-level tree is maintained in the master node to route the elements in the first levels where the partitioning is still coarse-grained. Once the elements have been mapped to the leaves, local subtrees perform the searches in parallel. The main idea behind here is that all trees act like a single metric tree but in a fully distributed way.

In order to remove the need for backtracking in M-tree or the use of redundancy, DS-RNGE implements an approximate approach that allows errors near borders in exchange for an improvement on efficiency. Authors argue that when the number of elements is much greater than the number of partitions, the number of conflicts becomes negligible.

DS-RNGE proceeds in two phases for each newly arrived batch of data: an edition/update phase which maintains and enhances the case-base, and a prediction phase that classifies new unlabeled data. Both phases require fast neighbor queries to accomplish their mission. To deal with this problem, authors propose a smart partitioning process in which each subtree queries only a single space partition.

In the following sections, we present the different procedures involved in DS-RNGE. Firstly, we describe the first steps to initialize the distributed case-base. Afterwards, the editing/updating process is presented. Here, we present details of insertion and deletion of examples in the tree. Finally, we describe the prediction phase in the last section.

5.5.1   Partitioning

The first step in DS-RNGE consists of building a distributed metric tree formed by a top-tree in the master machine and a set of local trees in the slave machines. This distributed tree will be queried and updated during next iterations with new batches. From the first batch, a sample of nt instances is taken in order to construct the main tree. The sampled data should be small enough to fit in a single machine and should maximize the separability between examples to avoid overlapping in the future subtrees. The routing tree is created following the standard procedure presented in [29], where upper and lower bounds are defined to control the size of nodes.

Once the top-tree is initialized, it is replicated to each machine and one subtree per leaf is created in the slave nodes. Then, every element in the first batch is inserted in the subtrees following these steps:

- For each element, the algorithm searches the nearest leaf node in the top-tree. According to the correspondence between leaf nodes and subtrees we can determine which subtree each element will be sent to. This process is performed in a map phase.

Algorithm 1 Initial partitioning process

Input: data, nt

 1: *// data is the input dataset*
 2: *// nt Number of leaf trees to be distributed across the nodes*
 3: *sample* = smartSampling(*data*)
 4: *topTree* = In the master machine, build the top M-tree using *sample* and the standard partitioning procedure showed in [29]. It will be replicated to every slave machine.
 5: For each leaf node in the *topTree*, one subtree is created in a single slave machine. The resulting set of trees (stored as an RDD) is partitioned and cached for further processing.
 6: **mapReduce** *e ∈ data*
 7:      Find the nearest leaf node to *e* in *topTree*, and outputs a tuple with the tree's ID (key) and *e* (value). (MAP)
 8:      The tuple is sent to the correspondent partition and attached to the subtree according to its key. (SHUFFLE)
 9:      Combine all the elements with the same key (tree ID) by inserting them into the local tree. (REDUCE)
10:      Return the updated tree.
11: **end mapReduce**

- The elements are shuffled to the subtrees according to their keys. Each subtree gets a list of elements to be inserted.
- For each subtree, all received elements are locally inserted in this tree. This process is performed in a reduce phase.

Note that the partitions/subtrees derived from this phase will be maintained during the complete process for re-usability purposes, so that only the arriving instances will be moved across the network in each iteration. Algorithm 1 shows this procedure in detail using a MapReduce syntax.

5.5.2   Updating the Distributed Tree with Edition

Whenever a new batch of data arrives to the system, the updating process is launched. This process inserts new correct examples, as well as removes those redundant examples already inserted. At first, the algorithm decides which subtree each element falls into following the same process described in the previous section. Once all instances are shuffled to the subtrees, a local nearest neighbor search for each element is started in the corresponding subtrees.

After all neighbors are collected, the IS algorithm creates groups where each group is formed by a new element and its neighbors. Then, local RNGE is applied on each group. The idea behind that is to build a local graph around each group and through this graph to decide what kind of action to perform on each element. New examples can be inserted or not, whereas old examples (neighbors) can be removed or conserved. Since each graph only has a narrow view of the case-base, the set of neighbors that can be removed is limited to those that share an edge with the new element.

Algorithm 2 Updating process with edition

Input: query, ks, ro
1: // *query is the data to be queried*
2: // *ks represents the number of neighbors to use in the IS phase.*
3: // *ro indicates whether to remove old noisy examples or not.*
4: **mapReduce** $e \in data$
5:     Find the nearest leaf node to e in $topTree$ and outputs a tuple with the tree's ID (key) and e
       (value). (MAP)
6:     The tuple is sent to the correspondent subtree according to its key. (SHUFFLE)
7:     $neighbors$ = the standard M-tree search process is launched for each element in its local
       subtree in order to retrieve the ks-neighbors of e. The output will consist of a tuple with e (key)
       and a list of its ks-neighbors (value). (REDUCE)
8:     edited = compute the local RNGE graph using e and ne, remove those elements that disagree
       with their neighbors.
9:     **if** $ro == true$ **then**
10:        Removed old noisy instances in $edited$ from the tree.
11:    **end if**
12:    Add new correct instances in $edited$ to the tree.
13:    Return the updated tree.
14: **end mapReduce**

Once decisions for each element are taken we perform insertions and removals locally in the subtrees in the own reduce phase. Notice that by doing so the neighbor query and the editing process are both performed in the same MapReduce process, thus reducing the communication overhead. The complete editing process is described in Algorithm 2.

Within the edition process, local construction of graphs and subsequent filtering is illustrated in Fig. 5.4. Here a new example E_{new} from class A (dashed circle) arrives to a given partition (Algorithm 2). From that partition, the $ks = 7$-NN are retrieved to build the graph shown in step 2. The left part in this figure represents how the graph looked before the arrival of E_{new}. Then, the graph is reconstructed following the rules defined by the relative neighborhood graph. As only two examples share edges with E_{new}, its neighborhood is formed by these points. Finally, removal decisions are made according to the connections between neighbors. In this case, E_{new} is prepared to be inserted in the case-base since all its edge-neighbors agree with its class. Red crosses can be also removed as they disagree with their neighbors.

5.5.3   Prediction

The labeling process is an approximate function started whenever new unlabeled data arrive at the system (see Algorithm 3). For each element the algorithm searches for the nearest leaf node in the master node and shuffles the elements to the correspondent subtrees. Next, the local search process is used to retrieve the kp-

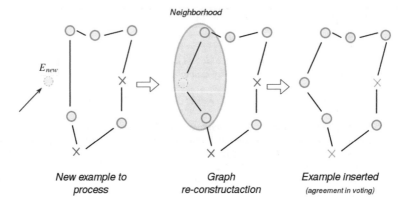

Fig. 5.4 Local graph edition for incoming instances. Class A instances are depicted as red crosses, and class B instances as blue circles. E_{new} is the new example to be processed. The result is the insertion of E_{new} and the removal of the class B examples

---

Algorithm 3 Prediction process
***

Input: query, kp
 1: // *query is the data to be queried*
 2: // *kp represents the number of neighbors for predictions.*
 3: **mapReduce** $e \in data$
 4:    Find the nearest leaf node to e in $topTree$ and outputs a tuple with the tree's ID (key) and e (value). (MAP)
 5:    The tuple is sent to the correspondent subtree according to its key. (SHUFFLE)
 6:    *neighbors* = the standard M-tree search process is launched for each element in its local subtree in order to retrieve the ks-neighbors of e. The output will consist of a tuple with e (key) and a list of its ks-neighbors (value). (REDUCE)
 7:    For each tuple in *neighbors* return the most-voted class from the list of neighbors. This value will be the class predicted for the given element.
 8: **end mapReduce**

---

neighbors of each new element. For each group, formed by a new element and its neighbors, the algorithm predicts the class for each new element by applying the majority voting scheme. Notice that the query and the prediction phases are both performed in the same MapReduce phase just like in the edition algorithm.

5.6 Summary and Conclusions

Data reduction can be seen as the set of techniques devoted to reduce the number of instances in the data while retaining as much information as possible. This alleviates both computing times and storage requirements. However, in Big Data scenarios, classic data reduction techniques cannot be applied.

This chapter introduces a novel framework for performing classic data reduction methods in Big Data domains. This framework is employed for the implementation of some classic data reduction algorithms, showing good results in both reduction rates and accuracy. Finally, a data reduction method for data streaming based on RNGE is presented.

Data redundancy seems to be a key issue in most of the analyzed datasets in the literature. Transforming these big amounts of information into smaller datasets heavily reduces the data storage requirements and the time needed to perform high-quality DM. There are many open challenges for the instance reduction problem in Big Data scenarios. The proposal of ensemble learning techniques based on diverse subsets of the data may improve the performance of current proposals. On the other hand, ad hoc instance reduction algorithms will be needed for specific DM algorithms to do a more tailored Smart Data reduction.

References

1. Angiulli, F. (2007). Fast nearest neighbor condensation for large data sets classification. *IEEE Transactions on Knowledge and Data Engineering, 19*(11), 1450–1464.
2. Arnaiz-González, Á., Díez-Pastor, J.-F., Rodríguez, J. J., & García-Osorio, C. (2016). Instance selection of linear complexity for big data. *Knowledge-Based Systems, 107*, 83–95.
3. Arnaiz-González, Á., González-Rogel, A., Díez-Pastor, J.-F., & López-Nozal, C. (2017). MR-DIS: Democratic instance selection for big data by MapReduce. *Progress in Artificial Intelligence, 6*, 1–9.
4. Bezdek, J. C., & Kuncheva, L. I. (2001). Nearest prototype classifier designs: An experimental study. *International Journal of Intelligent Systems, 16*(12), 1445–1473.
5. Cano, J. R., Herrera, F., & Lozano, M. (2003). Using evolutionary algorithms as instance selection for data reduction in KDD: An experimental study. *IEEE Transactions on Evolutionary Computation, 7*(6), 561–575.
6. Cano, J. R., Herrera, F., & Lozano, M. (2005). Stratification for scaling up evolutionary prototype selection. *Pattern Recognition Letters, 26*(7), 953–963.
7. Cano, J. R., Herrera, F., & Lozano, M. (2006). On the combination of evolutionary algorithms and stratified strategies for training set selection in data mining. *Applied Soft Computing, 6*(3), 323–332.
8. Chang, C. L. (1974). Finding prototypes for nearest neighbor classifiers. *IEEE Transactions on Computers, 100*(11), 1179–1184.
9. Chu, C.-T., Kim, S. K., Lin, Y.-A., Yu, Y. Y., Bradski, G., Ng, A. Y., et al. (2006). Map-reduce for machine learning on multicore. In *Proceedings of the 19th International Conference on Neural Information Processing Systems*, NIPS'06 (pp. 281–288).
10. Cover, T., & Hart, P. (1967). Nearest neighbor pattern classification. *IEEE Transactions on Information Theory, 13*(1), 21–27.
11. de Haro-García, A., & García-Pedrajas, N. (2009). A divide-and-conquer recursive approach for scaling up instance selection algorithms. *Data Mining and Knowledge Discovery, 18*(3), 392–418.
12. Derrac, J., García, S., & Herrera, F. (2010). IFS-CoCo: Instance and feature selection based on cooperative coevolution with nearest neighbor rule. *Pattern Recognition, 43*(6), 2082–2105.
13. Derrac, J., García, S., & Herrera, F. (2010). Stratified prototype selection based on a steady-state memetic algorithm: A study of scalability. *Memetic Computing, 2*(3), 183–199.
14. Dua, D., & Graff, C. (2017). *UCI machine learning repository*. http://archive.ics.uci.edu/ml

15. Eiben, A. E., & Smith, J. E. (2003). *Introduction to evolutionary computing* (Vol. 53). Berlin: Springer.
16. Gama, J., Ganguly, A., Omitaomu, O., Vatsavai, R., & Gaber, M. (2009). Knowledge discovery from data streams. *Intelligent Data Analysis, 13*(3), 403–404.
17. García, S., Cano, J. R., & Herrera, F. (2008). A memetic algorithm for evolutionary prototype selection: A scaling up approach. *Pattern Recognition, 41*(8), 2693–2709.
18. García, S., Derrac, J., Cano, J. R., & Herrera, F. (2012). Prototype selection for nearest neighbor classification: Taxonomy and empirical study. *IEEE Transactions on Pattern Analysis and Machine Intelligence, 34*(3), 417–435.
19. García, S., Luengo, J., & Herrera, F. (2014). *Data preprocessing in data mining*. Berlin: Springer Publishing Company, Incorporated.
20. García-Osorio, C., de Haro-García, A., & García-Pedrajas, N. (2010). Democratic instance selection: A linear complexity instance selection algorithm based on classifier ensemble concepts. *Artificial Intelligence, 174*(5), 410–441.
21. García-Pedrajas, N., & de Haro-García, A. (2012). Scaling up data mining algorithms: review and taxonomy. *Progress in Artificial Intelligence, 1*(1), 71–87.
22. García-Pedrajas, N., de Haro-García, A., & Pérez-Rodríguez, J. (2013). A scalable approach to simultaneous evolutionary instance and feature selection. *Information Sciences, 228*, 150–174.
23. Han, D., Giraud-Carrier, C., & Li, S. (2015). Efficient mining of high-speed uncertain data streams. *Applied Intelligence, 43*(4), 773–785.
24. Hart, P. E. (1968). The condensed nearest neighbor rule. *IEEE Transactions on Information Theory, 18*, 515–516.
25. Iguyon, I., & Elisseeff, A. (2003). An introduction to variable and feature selection. *Journal of Machine Learning Research, 3*, 1157–1182.
26. Kononenko, I. (1994). Estimating attributes: Analysis and extensions of relief. In *Machine Learning: European Conference on Machine Learning ECML-94* (pp. 171–182). Berlin: Springer.
27. Lam, W., Keung, C.-K., & Liu, D. (2002). Discovering useful concept prototypes for classification based on filtering and abstraction. *IEEE Transactions on Pattern Analysis and Machine Intelligence, 24*(8), 1075–1090.
28. Liu, H., & Motoda, H. (2007). *Computational methods of feature selection*. Boca Raton: CRC Press.
29. Liu, T., Moore, A. W., Gray, A. G., & Yang, K. (2004). An investigation of practical approximate nearest neighbor algorithms. In *NIPS'04 Proceedings of the 17th International Conference on Advances in Neural Information Processing Systems NIPS* (pp. 825–832).
30. Maillo, J., Ramírez, S., Triguero, I., & Herrera, F. (2017). kNN-IS: An Iterative Spark-based design of the k-Nearest Neighbors classifier for big data. *Knowledge-Based Systems, 117*, 3–15.
31. Navot, A., Shpigelman, L., Tishby, N., & Vaadia, E. (2006). Nearest neighbor based feature selection for regression and its application to neural activity. In *Proceedings of the International Conference on Advances in Neural Information Processing Systems* (pp. 996–1002).
32. Palma-Mendoza, R. J., Rodriguez, D., & de-Marcos, L. (2018). Distributed reliefF-based feature selection in spark. *Knowledge and Information Systems, 57*, 1–20.
33. Peralta, D., del Río, S., Ramírez-Gallego, S., Triguero, I., Benitez, J. M., & Herrera, F. (2016). Evolutionary feature selection for big data classification: A MapReduce approach. *Mathematical Problems in Engineering, 2015*, 1–11, Article ID 246139
34. Provost, F., & Kolluri, V. (1999). A survey of methods for scaling up inductive algorithms. *Data Mining and Knowledge Discovery, 3*(2), 131–169.
35. Ramírez-Gallego, S., Krawczyk, B., García, S., Woźniak, M., Benítez, J. M., & Herrera, F. (2017). Nearest neighbor classification for high-speed big data streams using spark. *IEEE Transactions on Systems, Man, and Cybernetics: Systems, 47*(10), 2727–2739.
36. Sánchez, J. S., Pla, F., & Ferri, F. J. (1997). Prototype selection for the nearest neighbour rule through proximity graphs. *Pattern Recognition Letters, 18*(6), 507–513.

37. Skalak, D. B. (1994). Prototype and feature selection by sampling and random mutation hill climbing algorithms. In *11th International Conference on Machine Learning (ML'94)* (pp. 293–301).
38. Triguero, I., Derrac, J., Garcia, S., & Herrera, F. (2012). A taxonomy and experimental study on prototype generation for nearest neighbor classification. *IEEE Transactions on Systems, Man, and Cybernetics, Part C (Applications and Reviews), 42*(1), 86–100.
39. Triguero, I., García, S., & Herrera, F. (2010). IPADE: Iterative prototype adjustment for nearest neighbor classification. *IEEE Transactions on Neural Networks, 21*(12), 1984–1990.
40. Triguero, I., García, S., & Herrera, F. (2011). Differential evolution for optimizing the positioning of prototypes in nearest neighbor classification. *Pattern Recognition, 44*(4), 901–916.
41. Triguero, I., García-Gil, D., Maillo, J., Luengo, J., García, S., & Herrera, F. (2019). Transforming big data into smart data: An insight on the use of the k-nearest neighbors algorithm to obtain quality data. *Wiley Interdisciplinary Reviews: Data Mining and Knowledge Discovery, 9*(2), e1289.
42. Triguero, I., Gonzalez, S., Moyano, J. M., García, S., Alcala-Fdez, J., Luengo, J., et al. (2017). Keel 3.0: An open source software for multi-stage analysis in data mining. *International Journal of Computational Intelligence Systems, 10*, 1238–1249.
43. Triguero, I., Peralta, D., Bacardit, J., García, S., & Herrera, F. (2015). MRPR: A MapReduce solution for prototype reduction in big data classification. *Neurocomputing, 150*, Part A, 331–345, 2015.
44. Wilson, D. L. (1972). Asymptotic properties of nearest neighbor rules using edited data. *IEEE Transactions on Systems, Man, and Cybernetics, 2*(3), 408–421.
45. Xue, B., Zhang, M., Browne, W. N., & Yao, X. (2016). A survey on evolutionary computation approaches to feature selection. *IEEE Transactions on Evolutionary Computation, 20*(4), 606–626.

Chapter 6
Imperfect Big Data

6.1 Introduction

Albeit most techniques and algorithms assume that the data is accurate, measurements in our analogical world are far from being perfect [8]. The alterations of the measured values can be caused by noise, an external process that generates corruption in the stored data, either by faults in data acquisition, transmission, storage, integration, and categorization [31]. The impact of noise in data has drawn the attention of researchers in the specialized literature [9]. The presence of noise has a severe impact in learning problems: to cope with the noise bias, the generated models are more complex, showing less generalization abilities, lower precision, and higher computational cost [36, 37].

Alleviating or removing the effects of noise implies that we need to identify the components in the data that are prone to be affected. The specialized literature often distinguishes between noise in the input variables (namely *attribute noise*) and the noise that affects the supervised features. Attribute noise may be caused by erroneous attribute values, MV and "do not care" values. Note that only in the case of supervised problems the noise in the output variables can exist. In classification, this kind of noise is often known either as *class* or *label noise*. The latter refers to instances belonging to the incorrect class either by contradictory examples [14] or misclassifications [37], due to labeling process subjectivity, data entry errors, or inadequacy of the information used to label each instance. In regression problems, noise in the output will appear as a bias added to the actual output value, resulting in a superposition of two different functions that it is difficult to separate.

MV, among all the corruptions in input attribute values, deserve special attention. In spite of being easily identifiable, MV pose a more severe impact in learning models, as most of the techniques assume that the training data provided is complete [13]. Until recently, practitioners opted to discard the examples containing MV, but this praxis often leads to severe bias in the inference process [19]. In fact, inappropriate MV handling will lead to model bias due to the distribution difference

© Springer Nature Switzerland AG 2020
J. Luengo et al., *Big Data Preprocessing*,
https://doi.org/10.1007/978-3-030-39105-8_6

among complete and incomplete data unless the MV are appropriately treated. Statistical procedures have been developed to impute (fill in) the MV to generate a complete dataset, obeying the underlying distributions in the data. The usage of machine learning approaches to perform imputation, as regressors or classifiers, quickly followed in the specialized literature, resulting in a large set of techniques that can be applied to cope with MV in the data [20].

The applicability of noise filters or MV imputations cannot be blindly carried out. The statistical dependencies among the corrupted and clean data will dictate how the imperfect data can be handled. Originally, Little and Rubin [19] described the three main mechanisms of MV introduction. When the MV distribution is independent of any other variable, we face missing completely at random (MCAR) mechanism. A more general case is when the MV appearance is influenced by other observed variables, constituting the missing at random (MAR) case. These two scenarios enable the practitioner to utilize imputators to deal with MV. Inspired by this classification, Frénay and Verleysen [8] extended this classification to noise data, analogously defining Noisy Completely at Random and Noisy at Random. Thus, methods that correct noise, as noise filters, can only be safely applied with these two scenarios as well.

Alternatively, the value of the attribute itself can influence the probability of having a MV or a noisy value. These cases were named as missing not at random (MNAR) and noisy not at random for MV and noisy data, respectively. Blindly applying imputators or noise correctors in this case will result in a data bias. In these scenarios, we need to model the probability distribution of the noisy or missingness mechanism by using expert knowledge and introduce it in statistical techniques as multiple imputation [23]. To avoid improperly application of correcting techniques, some test have been developed to evaluate the underlying mechanisms [18] but still careful data exploration must be carried out first.

In the rest of this chapter, we will analyze in depth the problem of noise (Sect. 6.2). Afterwards we shall outline the proposals developed until now to deal with it (Sects. 6.3 and 6.4). Section 6.5 is devoted to present the proposals for dealing with MV in Big Data. Finally, Sect. 6.6 summarizes the chapter and gives some conclusions.

6.2   Noise Filtering

In a classification problem, several effects of this noise can be observed by analyzing its spatial characteristics: noise may create small clusters of instances of a particular class in the instance space corresponding to another class, displace or remove instances located in key areas within a concrete class, or disrupt the boundaries of the classes resulting in an increased boundaries overlap. All these imperfections may harm data interpretation, the design, size, building time, interpretability, and accuracy of models, as well as decision-making [37].

As described by Wang et al. [35], from the large number of components that comprise a dataset, class labels and attribute values are two essential elements in classification datasets. Thus, two types of noise are commonly differentiated in the literature [35, 37]:

- *Class noise*, also known as *label noise*, takes place when an example is wrongly labeled. Class noise includes contradictory examples [24] (examples with identical input attribute values having different class labels) and misclassifications [37] (examples which are incorrectly labeled).
- *Attribute noise* refers to corruptions in the values of the input attributes. It includes erroneous attribute values, MV and incomplete attributes or "do not care" values. MV are usually considered independently in the literature, so *attribute noise* is mainly used for erroneous values [37].

Class noise is generally considered more harmful to the learning process, and methods for dealing with class noise are more frequent in the literature [37]. Class noise may have many reasons, such as errors or subjectivity in the data labeling process, as well as the use of inadequate information for labeling. Data labeling by domain experts is generally costly, and automatic taggers are used, increasing the probability of class noise.

Due to the increasing attention from researchers and practitioners, numerous techniques have been developed to tackle it [8, 11, 37]. These techniques include learning algorithms robust to noise as well as data preprocessing techniques that remove or "repair" noisy instances. In [8] the mechanisms that generate label noise are examined, relating them to the appropriate treatment procedures that can be safely applied:

- On the one hand, *algorithm level* approaches attempt to create robust classification algorithms that are little influenced by the presence of noise. This includes approaches where existing algorithms are modified to cope with label noise by either being modeled in the classifier construction [17], by applying pruning strategies to avoid overfitting or by diminishing the importance of noisy instances with respect to clean ones [22]. Recent proposals exist which that combine these two approaches, which model the noise and give less relevance to potentially noisy instances in the classifier building process [3].
- On the other hand, *data level* approaches (also called *filters*) try to develop strategies to cleanse the dataset as a previous step to the fit of the classifier, by either creating ensembles of classifiers [4], partitioning the data [33], iteratively filtering noisy instances [15], computing metrics on the data or even hybrid approaches that combine several of these strategies.

In the Big Data environment there is a special need for noise filter methods. It is well known that the high dimensionality and example size generate accumulated noise in Big Data problems [7]. Noise filters reduce the size of the datasets and improve the quality of the data by removing noisy instances, but most of the classic algorithms for noisy data, noise filters in particular, are not prepared for working with huge volumes of data as they have an iterative approach. In the following two sections (Sects. 6.3 and 6.4) are devoted to outline the most relevant proposals until now to deal with noise in Big Data classification.

6.3   Enabling Smart Data: Noise Filtering in Big Data Classification

In this section, we will describe a framework composed of three algorithms, focused on noise filtering in Big Data classification [12]. This paper presents the first suitable noise filters in Big Data domains, where the high redundancy of the instances and high dimensional problems pose new challenges to classic noise preprocessing algorithms. Authors propose a framework for Big Data under Apache Spark for removing noisy examples composed of two algorithms based on ensembles of classifiers. The first one is a homogeneous ensemble, named Homogeneous Ensemble for Big Data (HME-BD), which uses a single base classifier (Random Forest) over a partitioning of the training set. The second ensemble is a heterogeneous ensemble, namely Heterogeneous Ensemble for Big Data (HTE-BD), that uses different classifiers to identify noisy instances: Random Forest, logistic regression, and K-nearest neighbors (KNN) as base classifiers. Authors also considered a simple filtering approach based on similarities between instances, named Edited Nearest Neighbor for Big Data (ENN-BD). ENN-BD examines the nearest neighbors of every example in the training set and eliminates those whose majority of neighbors belong to a different class. In Fig. 6.1 we can find a Spark package associated with this research.

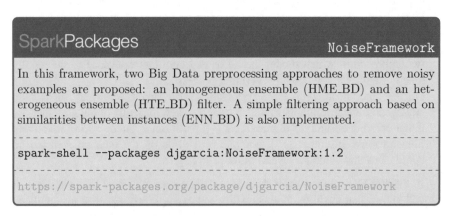

Fig. 6.1 Spark package: NoiseFramework

6.3.1   HME-BD: Homogeneous Ensemble

The homogeneous ensemble is inspired by Cross-Validated Committees Filter (CVCF) [33]. This filter removes noisy examples by partitioning the data in P subsets of equal size. Then, a decision tree, such as C4.5, is learned P times, each time leaving out one of the subsets of the training data. This results in P classifiers which are used to predict all the training data P times. Then, using a voting strategy, misclassified instances are removed.

HME-BD is also based on a partitioning scheme of the training data. There is an important difference with respect to CVCF: the use of Spark's implementation of Random Forest instead a of a decision tree as a classifier. CVCF creates an ensemble from partitioning of the training data. HME-BD also partitions the training data, but the use of Random Forest allows us to improve the voting step:

- CVCF predicts the whole dataset P times. HME-BD only predicts the instances of the partition that Random Forest has not seen while learning the model. This step is repeated P times. With this change it not only improves the performance, but also the computing time of the algorithm since it only has to predict a small part of the training data each iteration.
- HME-BD does not need to implement a voting strategy, the decision of whether an instance is noisy is associated with the Random Forest prediction.

Algorithm 1 describes the noise filtering process in HME-BD:

Algorithm 1 HME-BD algorithm

Input: *data* a RDD of tuples (label, features)
Input: P the number of partitions
Input: *nTrees* the number of trees for Random Forest
Output: the filtered RDD without noise
 1: $partitions \leftarrow kFold(data, P)$
 2: $filteredData \leftarrow \emptyset$
 3: **for all** $train, test \in partitions$ **do**
 4:    $rfModel \leftarrow randomForest(train, nTrees)$
 5:    $rfPred \leftarrow predict(rfModel, test)$
 6:    $joinedData \leftarrow join(zipWithIndex(test), zipWithIndex(rfPred))$
 7:    $markedData \leftarrow$
 8:    **map** $original, prediction \in joinedData$
 9:      **if** $label(original) = label(prediction)$ **then**
10:         $original$
11:      **else**
12:         $(label = \emptyset, features(original))$
13:      **end if**
14:    **end map**
15:    $filteredData \leftarrow union(filteredData, markedData)$
16: **end for**
17: **return** $(filter(filteredData, label \neq \emptyset))$

- The algorithm filters the noise in a dataset by performing a $kFold$ on the training data. Spark's $kFold$ function returns a list of $(train, test)$ for a given P, where $test$ is a unique 1/kth of the data, and $train$ is a complement of the $test$ data.
- It iterates through each partition, learning a Random Forest model using the $train$ as input data and predicting the $test$ using the learned model.
- In order to join the $test$ data and the predicted data for comparing the classes, it uses the $zipWithIndex$ operation in both RDD. With this operation, an index is added to each element of both RDD. This index is used as key for the join operation.
- The next step is to apply a map function to the previous RDD in order to check for each instance the original class and the predicted one. If the predicted class and the original are different, the instance is marked as noise.
- The result of the previous map function is a RDD where noisy instances are marked. These instances are finally removed using a $filter$ function and the resulting dataset is returned.

In Fig. 6.2 we can see a flowchart of the HME-BD noise filtering process. In this figure we can see how data is partitioned and learned using the $train$ partition, then the $test$ partition is predicted using the model learned. Finally, wrongly predicted instances are removed and the partitions are joined together.

6.3.2 HTE-BD: Heterogeneous Ensemble

Heterogeneous ensemble is inspired by ensemble filter (EF) [4]. This noise filter uses a set of three learning algorithms for identifying mislabeled instances in a dataset: a univariate decision tree (C4.5), KNN, and a linear machine. It performs a k-fold cross validation over the training data. For each one of the k parts, three algorithms are trained on the other $k - 1$ parts. Each of the classifiers is used to tag each of the $test$ examples as noisy or clean. At the end of the k-fold, each example of the input data has been tagged. Finally, using a voting strategy, a decision is made and noisy examples are removed.

HTE-BD follows the same working scheme as EF. The main difference is the choice of the three learning algorithms:

- Instead of a decision tree, HTE-BD uses Spark's implementation of Random Forest.
- It uses an exact implementation of KNN with the Euclidean distance present in Spark's community repository, KNN-IS [21].
- The linear machine has been replaced by Spark's implementation of logistic regression, which is another linear classifier.

The noise filtering process in HTE-BD is shown in Algorithm 2:

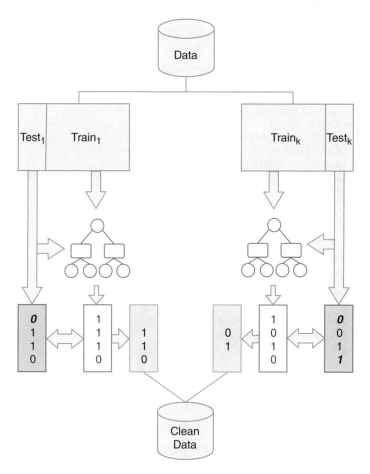

Fig. 6.2 HME-BD noise filtering process flowchart

- For each *train* and *test* partition of the *k*-fold performed to the input data, it learns three classification algorithms: Random Forest, logistic regression, and 1NN using the *train* as input data.
- Then it predicts the *test* data using the three learned models. This creates a RDD of triplets (rf, lr, knn) with the prediction of each algorithm for each instance.
- The predictions and the *test* data are joined by index in order to compare the predictions and the original label.
- It compares the three predictions of each instance in the *test* data with the original label using a map function and, depending upon the voting strategy, the instance is marked as noise or clean.
- Once the map function has been applied to each instance, noisy data is removed using a *filter* function and the dataset is returned.

Algorithm 2 HTE-BD Algorithm

Input: *data* a RDD of tuples (label, features)
Input: *P* the number of partitions
Input: *nTrees* the number of trees for Random Forest
Input: *vote* the voting strategy (majority or consensus)
Output: the filtered RDD without noise
  1: $partitions \leftarrow kFold(data, P)$
  2: $filteredData \leftarrow \emptyset$
  3: **for all** $train, test$ in $partitions$ **do**
  4:     $classifiersModel \leftarrow learnClassifiers(train, nTrees)$
  5:     $predictions \leftarrow predict(classifiersModel, test)$
  6:     $joinedData \leftarrow join(zipWithIndex(predictions), zipWithIndex(test))$
  7:     $markedData \leftarrow$
  8:     **map** $rf, lr, knn, orig \in joinedData$
  9:         $count \leftarrow 0$
 10:         **if** $rf \neq label(orig)$ **then** $count \leftarrow count + 1$ **end if**
 11:         **if** $lr \neq label(orig)$ **then** $count \leftarrow count + 1$ **end if**
 12:         **if** $knn \neq label(orig)$ **then** $count \leftarrow count + 1$ **end if**
 13:         **if** $vote = majority$ **then**
 14:             **if** $count \geq 2$ **then** $(label = \emptyset, features(orig))$ **end if**
 15:             **if** $count < 2$ **then** $orig$ **end if**
 16:         **else**
 17:             **if** $count = 3$ **then** $(label = \emptyset, features(orig))$ **end if**
 18:             **if** $count \neq 3$ **then** $orig$ **end if**
 19:         **end if**
 20:     **end map**
 21:     $filteredData \leftarrow union(filteredData, markedData)$
 22: **end for**
 23: **return** $(filter(filteredData, label \neq \emptyset))$

In Fig. 6.3 we show a flowchart of the HTE-BD noise filtering process. In this figure, the partitioning process is depicted. We can also observe the learning and prediction phase using the three classifiers, and the use of the voting strategy.

6.3.3   ENN-BD: Similarity Based Method

ENN-BD is a simple filtering algorithm based on KNN. It has been designed based on the edited nearest neighbor algorithm (ENN) [34] and follows a similarity between instances approach. ENN removes noisy instances in a dataset by comparing the label of each example with its closest neighbor. If the labels are different, the instance is considered as noisy and removed.

ENN-BD performs a 1-NN using Spark's community repository KNN-IS with the Euclidean distance. It checks for each instance if its closest neighbor belongs to the same class. In case the classes are different, the instance is marked as noise. Finally, marked instances are removed from the training data. This process is described in Algorithm 3.

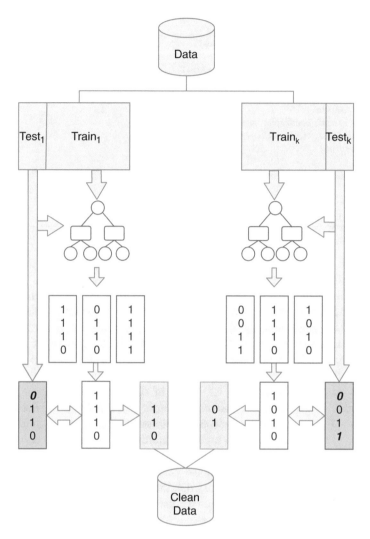

Fig. 6.3 HTE-BD noise filtering process flowchart

6.3.4 Noise Filtering: An Experimental Study

This section is devoted to show the performance of the above three noise filtering methods over four large-scale problems. Additionally, this section aims at showing that noise filtering is useful in Big Data classification, providing an improvement in classification accuracy in the presence of noise.

For such purpose, the three noise filtering methods are analyzed using four Big Data problems. These datasets have very different properties among them. SUSY dataset consists of 5,000,000 instances and 18 attributes [6]. The first eight

---

Algorithm 3 ENN-BD algorithm

Input: *data* a RDD of tuples (label, features)
Output: the filtered RDD without noise
  1: $knnModel \leftarrow KNN(1, "euclidean", data)$
  2: $knnPred \leftarrow zipWithIndex(predict(knnModel, data))$
  3: $joinedData \leftarrow join(zipWithIndex(data), knnPred)$
  4: $filteredData \leftarrow$
  5: **map** $original, prediction \in joinedData$
  6:   **if** $label(original) = label(prediction)$ **then**
  7:     $original$
  8:   **else**
  9:     $(noise, features(original))$
 10:   **end if**
 11: **end map**
 12: **return** $(filter(filteredData, label \neq noise))$

---

features are kinematic properties measured by the particle detectors at the Large Hadron Collider. The last ten are functions of the first eight features. The task is to distinguish between a signal process which produces supersymmetric (SUSY) particles and a background process which does not [1]. HIGGS dataset, which has 11,000,000 instances and 28 attributes [6]. This dataset is a classification problem to distinguish between a signal process which produces Higgs bosons and a background process which does not. Epsilon dataset, which consists of 500,000 instances with 2000 numerical features. This dataset was artificially created for the Pascal Large Scale Learning Challenge in 2008. It was further preprocessed and included in the LIBSVM dataset repository [5]. Finally, ECBDL14 dataset, which has 32 million instances and 631 attributes (including both numerical and categorical) [29]. This dataset was used as a reference at the machine learning (ML) competition of the Evolutionary Computation for Big Data and Big Learning held on July 14, 2014, under the international conference GECCO-2014. It is a binary classification problem where the class distribution is highly imbalanced: 98% of negative instances. For this problem, we use a reduced version with 1,000,000 instances and 30% of positive instances.

The experiments five levels of uniform class noise have been chosen: for each level of noise, a percentage of the training instances are altered by replacing their actual label by another label from the available classes. The selected noise levels are 0%, 5%, 10%, 15%, and 20%. In this case, a 0% noise level indicates that the dataset was unaltered. Noise has been introduced randomly to the class labels using a Spark Package called `RandomNoise`. It is a simple but useful package that adds class noise randomly into an RDD. In Fig. 6.4 we can find a Spark Package associated with this research.

For HME-BD and HTE-BD, 4 training partitions are selected. For HTE-BD, two voting strategies are used: *consensus* (same result for all classifiers) and *majority* (same result for at least half the classifiers). For ENN-BD, $k = 1$ is chosen.

SparkPackages   `RandomNoise`

This package adds class noise randomly into an RDD.

- -

`spark-shell --packages djgarcia:RandomNoise:1.0`

- -

`https://spark-packages.org/package/djgarcia/RandomNoise`

Fig. 6.4   Spark package: RandomNoise

In order to evaluate the effectiveness of the filtering proposals, one MLlib classifier, a decision tree, is used. Prediction accuracy is used to evaluate the model's performance produced by the classifiers (number of examples correctly labeled as belonging to a given class divided by the total number of elements).

Finally, the cluster used is composed of 20 computing nodes and one master node. The computing nodes hold the following characteristics: 2 processors x Intel(R) Xeon(R) CPU E5-2620, 6 cores per processor, 2.00 GHz, 2 TB HDD, 64 GB RAM. Regarding software, we have used the following configuration: Hadoop 2.6.0-cdh5.4.3 from Cloudera's open-source Apache Hadoop distribution, Apache Spark and MLlib 1.6.0, 460 cores (23 cores/node), 960 RAM GB (48 GB/node).

Table 6.1 gathers the test accuracy values for the three noise filter methods using a deep decision tree (*depth* = 20). The results presented have shown the importance of applying a noise treatment strategy, no matter how much noise is present in the dataset. The best performing filter overall is HME-BD, maintaining almost the same accuracy for every level of noise. For HTE-BD, the *consensus* strategy is performing better than the *majority* strategy due to higher instance removal. ENN-BD performance is behind the other two filters.

In Table 6.2 we present the reduction rate after the application of the three noise filtering methods for the four datasets. As we can expect, the higher the percentage of noise, the lower the number of instances that remain in the dataset after applying the filtering technique. In general, HME-BD is the most balanced technique in terms of instances removed and kept. Although the reduction rate by HTE-BD with majority voting is very similar to HME-BD, the instances selected to be eliminated are different, severely affecting the classifier used afterwards. ENN-BD is the filter that removes more instances. This aggressive filtering hinders the performance of noise tolerant classifiers, such as the decision tree.

In Table 6.3 we can see the average runtimes of the three methods for the four datasets in seconds. As the level of noise is not a factor that affects the runtime, we show the average of the five executions performed for each dataset. Measured times

Table 6.1 Decision tree test accuracy

Dataset				HTE-BD		
Vote	Noise (%)	Original	HME-BD	Majority	Consensus	ENN-BD
SUSY	0	80.24	79.78	79.69	**80.27**	78.56
	5	79.94	79.99	80.07	**80.36**	77.49
	10	79.15	79.85	79.81	**80.04**	77.00
	15	78.21	**79.81**	79.32	79.47	75.81
	20	77.09	**79.71**	79.35	78.95	74.21
HIGGS	0	70.17	**71.16**	69.61	70.41	68.85
	5	69.61	**71.14**	69.34	69.98	68.29
	10	69.22	**71.06**	68.95	69.56	67.52
	15	68.65	**71.03**	68.52	69.04	66.93
	20	67.82	**71.05**	68.18	68.38	66.05
Epsilon	0	62.39	**66.86**	65.13	66.07	61.54
	5	61.10	**66.64**	65.32	66.09	60.41
	10	60.09	**66.87**	65.46	66.11	59.20
	15	59.02	**66.62**	65.33	65.99	58.09
	20	57.73	**66.46**	65.08	65.69	56.71
ECBDL14	0	73.98	**74.59**	74.21	74.51	73.66
	5	72.87	**74.64**	74.16	74.54	73.48
	10	71.67	**74.59**	73.84	74.51	72.75
	15	70.28	**74.61**	73.82	73.91	71.68
	20	68.66	**74.83**	73.78	73.82	70.16

The highest accuracy value per dataset and noise level is stressed in bold

show that HME-BD is not only the best performing option in terms of accuracy, but also the most efficient one in terms of computing time. HME-BD is about ten times faster than the heterogeneous filter HTE-BD and the similarity filter ENN-BD. This is caused by the usage of the KNN classifier by HTE-BD and ENN-BD, which is very demanding in computing terms. As a result, HME-BD does not need to compute any distance measures, saving computing time and being the most recommended option to deal with noise in Big Data problems.

The extensive experimental study carried out in this section has shown that the usage of any of the noise treatment techniques in the framework always improves the *Original* accuracy value at the same noise level. HME-BD has shown to be the best performing method overall for the decision tree. It is also the most efficient method in terms of computing time. The voting strategy has a huge impact in the number of removed instances. As we could expect, KNN is a very demanding method in computing terms. This is reflected in the longer computing time of HTE-BD and ENN-BD.

Table 6.2 Reduction rate (%) for HME-BD, HTE-BD, and ENN-BD

Dataset			HTE-BD		
Vote	Noise (%)	HME-BD	Majority	Consensus	ENN-BD
SUSY	0	20.62	21.04	8.74	49.51
	5	23.57	25.09	10.33	49.57
	10	26.50	27.94	11.68	49.66
	15	29.44	30.85	13.04	49.74
	20	32.35	33.71	14.22	49.82
HIGGS	0	29.08	35.13	8.20	49.71
	5	31.15	36.65	8.83	49.75
	10	33.23	38.11	9.59	49.81
	15	35.38	39.55	10.35	49.92
	20	37.34	41.05	11.11	49.88
Epsilon	0	34.31	22.30	2.90	49.97
	5	25.32	25.24	4.32	50.01
	10	27.80	27.88	5.83	49.97
	15	30.79	30.71	7.22	50.01
	20	33.52	33.44	8.74	50.17
ECBDL14	0	22.44	21.35	5.85	26.58
	5	25.80	24.51	8.25	31.06
	10	28.49	27.68	10.31	35.07
	15	31.13	30.71	12.07	38.63
	20	33.86	33.69	13.91	41.60

Table 6.3 Average runtimes for HME-BD, HTE-BD, and ENN-BD in seconds

Dataset		HTE-BD		
Vote	HME-BD	Majority	Consensus	ENN-BD
SUSY	513.46	5511.15	5855.66	8956.71
HIGGS	587.72	15,300.62	15,232.99	25,441.09
Epsilon	1868.75	4120.79	7201.05	2718.97
ECBDL14	1228.24	9710.70	11,217.02	14,080.03

6.4   Noisy Big Data Treatment with the KNN Algorithm

In this section, we will describe a series of algorithms based on KNN for performing noise filtering [31]. As we have mentioned previously, the presence of noise involves a negative impact in the model obtained. This effect is aggravated if the learning technique is noise sensitive. In particular, the KNN algorithm is very sensitive to noise, especially when the value of k is low. KNN has been the seminal method to remove redundant and noisy instances in learning problems. The key idea of KNN, distance-based similarity, has been recurrently used to detect and remove class noise.

The literature in the usage of KNN to clean datasets is very prolific and span over several categories or topics. For instance, in [10] and [30], the authors categorized noise filtering techniques based on KNN as sub-families of prototype selection (PS) and prototype generation methods: edition-based methods and class-relabeling methods, respectively. The objective of edition-based methods is to only eliminate noisy instances (in contradistinction to more general PS methods that also remove redundant samples), and class-relabeling methods do not always remove the noisy instances, but they may amend those labels that the method found mistakenly assigned [25].

Among all the previous categories, one of the most popular methods is the edited nearest neighbor (ENN) [34], which removes all incorrectly labeled instances that do not agree with their k-nearest neighbors. If the labels are different, the instance is considered as noisy and removed. Other relevant examples of this family of methods are: All-KNN [28], NCN-Edit [25], or RNG [26]. A distributed version of the ENN algorithm based on Apache Spark can be found in Sect. 6.3.3 for very large datasets. This distributed version of ENN performs a global filtering of the instances, considering the whole dataset at once. The time complexity of this method is reduced to the same time complexity of the KNN.

In this work, authors have made the All-KNN algorithm global as ENN, as this algorithm basically consists of applying multiple times ENN. However, NCN-Edit and RNG methods have been considered within the MRPR framework proposed in [32] to make them scalable to Big Data. In Fig. 6.5 we can find this software available as a Spark Package.

6.5   Missing Values Imputation

In this section, we will describe the KNN-LI algorithm (k-Nearest Neighbors Local Imputation), focused on the missing values imputation in Big Data problem [31].

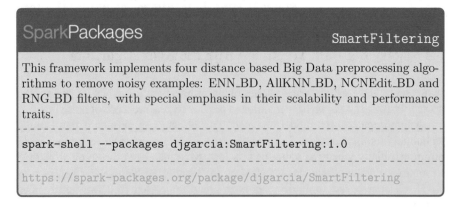

Fig. 6.5 Spark package: SmartFiltering

There are different ways to approach the problem of MV. For the sake of simplicity, we will focus on the MCAR and MAR cases by using imputation techniques, as MNAR will imply a particular solution and modeling for each problem. When facing MAR or MCAR scenarios, the simplest strategy is to discard those instances that contain MV. However, these instances may contain relevant information or the number of affected instances may also be extremely high, and therefore, the elimination of these samples may not be practical or even bias the data.

Instead of eliminating the corrupted instances, the imputation of MV is a popular option. The simplest and most popular estimate used to impute is the average value of the whole dataset, or the mode in case of categorical variables. Mean imputation would constitute a perfect candidate to be applied in Big Data environments as the mean of each variable remains unaltered and can be performed in $\mathcal{O}(n)$. However, this procedure presents drawbacks that discourage its usage: the relationship among the variables is not preserved and that is the property that learning algorithms want to exploit. Additionally, the standard error of any procedure applied to the data is severely underestimated [19] leading to incorrect conclusions.

Further developments in imputation are to solve the limitations of the two previous strategies. Statistical techniques such as expectation–maximization [27] or local least squares imputation [16] were applied in bioinformatics or climatic fields. Note that imputing MV can be described as a regression or classification problem, depending on the nature of the missing attribute. Shortly after, computer scientists propose the usage of ML algorithms to impute MV [20].

One of the most popular imputation approaches is based on KNN (denoted as KNN-I) [2]. In this algorithm, for each instance that contains one or more MV, it calculates the k-nearest neighbors and the gaps are imputed based on the existing values of the selected neighbors. If the value is nominal or categorical, it is imputed by the statistical mode. If the value is numeric, it will be imputed with the average of the nearest neighbors. A similarity function is used to obtain the k-nearest neighbors. The most commonly used similarity function for MV imputation is a variation of the Euclidean distance that accounts for those samples that contain MV. The advantage of KNN-I is that is both simple and flexible, requiring few parameters to operate and being able to use incomplete instances as neighbors. Most imputation algorithms only utilize complete instances to generate the imputation model, resulting in an approximate or biased estimation when the number of instances affected by MVs is high.

The proposal of imputation techniques in Big Data is still an open challenge, due to the difficulties associated to adapt complex algorithms to deal with partial folds of the data without losing predictive power. At this point, MV pose an important pitfall in the transition from Big to Smart Data. To the best of our knowledge, there has not been proposed a way of applying KNN-I on Big Data datasets. Although further investigation is required, authors propose a simple yet powerful approach to handle MV with the KNN-I algorithm on Big Data problems, which will be called k-Nearest Neighbors Local Imputation (KNN-LI). Figure 6.6 shows the workflow of the algorithm. Due to the scalability problems to tackle the Euclidean distance with

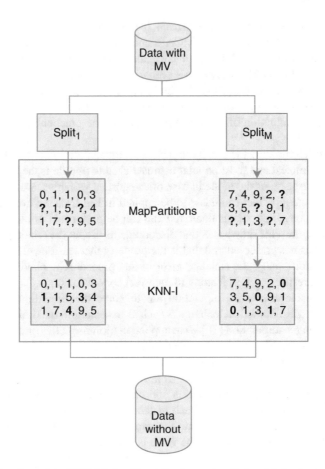

Fig. 6.6 Flowchart of the KNN-LI algorithm. The dataset is split into M chunks (map function) that are processed locally by a standard KNN-I algorithm. The resulting amended partitions are then gathered together

MV, the proposed KNN-LI algorithm follows a divide-and-conquer scheme under the MapReduce paradigm and it is implemented under the Apache Spark platform. It begins by splitting and distributing the dataset between the worker nodes. For each chunk of data, we compute the KNN-I method locally with the existing instances. Once all MV have been imputed for each chunk of the data, the results are simply grouped together to obtain a whole dataset free of MV. This local design is similar to the one followed in [32], for instance, reduction approaches, and allows us to impute MV in very large datasets. Nevertheless, as a local model we are aware that the quality of the imputation may vary depending on the number of partitions considered.

In Fig. 6.7 we can find this algorithm available publicly as a Spark Package in the Spark Packages repository.

SparkPackages **Smart_Imputation**

This contribution implements two approaches of the k Nearest Neighbor Imputation focused on the scalability in order to handle big dataset. k Nearest Neighbor - Local Imputation and k Nearest Neighbor Imputation - Global Imputation. The global proposal takes into account all the instances to calculate the k nearest neighbors. The local proposal considers those that are into the same partition, achieving higher times, but losing the information because it does not consider all the samples.

```
spark-shell --packages JMailloH:Smart_Imputation:1.0
```

https://spark-packages.org/package/JMailloH/Smart_Imputation

Fig. 6.7 Spark package: Smart_Imputation

6.6 Summary and Conclusions

In this chapter, a number of techniques for noise filtering and MV imputation are introduced. These techniques are used for amending what is called imperfect data. Noise affects the class labels, disrupting the boundaries of the problem and decreasing the performance of the classifiers. On the other hand, MV affects the learning process, as learners expect that the data is complete.

We have presented a number of algorithms for noise filtering. These algorithms are separated into two different proposals. The first one is based on ensembles of classifiers while the second uses the KNN algorithm underneath. Both have shown to obtain better results in noisy data than a no filtering strategy. Finally, a MV proposal is also presented. This method imputes the MV using the KNN algorithm with a divide-and-conquer scheme.

Nevertheless, the treatment of imperfect data in Big Data is far from being complete. The massive amount of data may result in data redundancy, leaving useless traditional approaches to imperfect data, as some regions can be densely populated and the elimination of corrupted examples can be safely done. However, the redundancy can be presented in an imbalanced way, thus requiring a careful study on which regions need to be carefully treated. Such an imbalance can pose a problem when the overlapping is high, as the boundaries can be difficult to estimate, hindering the algorithm's ability to detect noisy instances. Contradictory examples can also appear in numbers, without an easy approach. Imperfect data approaches that need to estimate the data density distribution are still a challenge, due to the computational demanding computations needed and the necessity of clean patches of examples in the instance space. In summary, while non-parametric techniques have been presented in this chapter to deal with imperfect data, its treatment is still an open but interesting challenge to be yet explored.

References

1. Baldi, P., Sadowski, P., & Whiteson, D. (2014). Searching for exotic particles in high-energy physics with deep learning. *Nature Communications, 5*, 4308.
2. Batista, G. E. A. P. A., & Monard, M. C. (2003). An analysis of four missing data treatment methods for supervised learning. *Applied Artificial Intelligence, 17*(5–6), 519–533.
3. Bouveyron, C., & Girard, S. (2009). Robust supervised classification with mixture models: Learning from data with uncertain labels. *Pattern Recognition, 42*(11), 2649–2658.
4. Brodley, C. E., & Friedl, M. A. (1999). Identifying mislabeled training data. *Journal of Artificial Intelligence Research, 11*, 131–167.
5. Chang, C.-C., & Lin, C.-J. (2011). LIBSVM: A library for support vector machines. *ACM Transactions on Intelligent Systems and Technology (TIST), 2*(3), 27:1–27:27.
6. Dua, D., & Graff, C. (2013). *UCI machine learning repository*. http://archive.ics.uci.edu/ml
7. Fan, J., Han, F., & Liu, H. (2014). Challenges of big data analysis. *National Science Review, 1*(2), 293–314.
8. Frénay, B., & Verleysen, M. (2014). Classification in the presence of label noise: A survey. *IEEE Transactions on Neural Networks and Learning Systems, 25*(5), 845–869.
9. Garcia, L. P. F., de Carvalho, A. C. P. L. F., & Lorena, A. C. (2015). Effect of label noise in the complexity of classification problems. *Neurocomputing, 160*, 108–119.
10. García, S., Derrac, J., Cano, J. R., & Herrera, F. (2012). Prototype selection for nearest neighbor classification: Taxonomy and empirical study. *IEEE Transactions on Pattern Analysis and Machine Intelligence, 34*(3), 417–435.
11. García, S., Luengo, J., & Herrera, F. (2015). *Data preprocessing in data mining*. Berlin: Springer.
12. García-Gil, D., Luengo, J., García, S., & Herrera, F. (2019). Enabling smart data: Noise filtering in big data classification. *Information Sciences, 479*, 135–152.
13. García-Laencina, P. J., Sancho-Gómez, J.-L., & Figueiras-Vidal, A. R. (2010). Pattern classification with missing data: A review. *Neural Computing and Applications, 19*(2), 263–282.
14. Hernández, M. A., & Stolfo, S. J. (1998). Real-world data is dirty: Data cleansing and the merge/purge problem. *Data Mining and Knowledge Discovery, 2*, 9–37.
15. Khoshgoftaar, T. M., & Rebours, P. (2007). Improving software quality prediction by noise filtering techniques. *Journal of Computer Science and Technology, 22*, 387–396.
16. Kim, H., Golub, G. H., & Park, H. (2004). Missing value estimation for DNA microarray gene expression data: local least squares imputation. *Bioinformatics, 21*(2), 187–198.
17. Li, Y., Wessels, L. F. A., de Ridder, D., & Reinders, M. J. T. (2007). Classification in the presence of class noise using a probabilistic Kernel Fisher method. *Pattern Recognition, 40*(12), 3349–3357.
18. Little, R. J. A. (1988). A test of missing completely at random for multivariate data with missing values. *Journal of the American Statistical Association, 83*(404), 1198–1202.
19. Little, R. J. A., & Rubin, D. B. (2014). *Statistical analysis with missing data* (Vol. 333). Hoboken: Wiley.
20. Luengo, J., García, S., & Herrera, F. (2012). On the choice of the best imputation methods for missing values considering three groups of classification methods. *Knowledge and Information Systems, 32*(1), 77–108.
21. Maíllo, J., Ramírez, S., Triguero, I., & Herrera, F. (2017). kNN-IS: An Iterative Spark-based design of the k-nearest neighbors classifier for big data. *Knowledge-Based Systems, 117*, 3–15.
22. Miao, Q., Cao, Y., Xia, G., Gong, M., Liu, J., & Song, J. (2016). RBoost: Label noise-robust boosting algorithm based on a nonconvex loss function and the numerically stable base learners. *IEEE Transactions on Neural Networks and Learning Systems, 27*(11), 2216–2228.
23. Royston, P. (2014). Multiple imputation of missing values. *Stata Journal, 4*(3), 227–41.
24. Sáez, J. A., Galar, M., Luengo, J., & Herrera, F. (2016). INFFC: An iterative class noise filter based on the fusion of classifiers with noise sensitivity control. *Information Fusion, 27*, 19–32.

25. Sánchez, J. S., Barandela, R., Marqués, A. I., Alejo, R., & Badenas, J. (2003). Analysis of new techniques to obtain quality training sets. *Pattern Recognition Letters, 24*(7), 1015–1022.
26. Sánchez, J. S., Pla, F., & Ferri, F. J. (1997). Prototype selection for the nearest neighbor rule through proximity graphs. *Pattern Recognition Letters, 18*, 507–513.
27. Schneider, T. (2001). Analysis of incomplete climate data: Estimation of mean values and covariance matrices and imputation of missing values. *Journal of Climate, 14*(5), 853–871.
28. Tomek, I. (1976). An experiment with the edited nearest-neighbor rule. *IEEE Transactions on Systems and Man and Cybernetics, 6*(6), 448–452.
29. Triguero, I., del Río, S., López, V., Bacardit, J., Benítez, J. M., & Herrera, F. (2015). ROSEFW-RF: the winner algorithm for the ecbdl14 big data competition: an extremely imbalanced big data bioinformatics problem. *Knowledge-Based Systems, 87*, 69–79.
30. Triguero, I., Derrac, J., García, S., & Herrera, F. (2012). A taxonomy and experimental study on prototype generation for nearest neighbor classification. *IEEE Transactions on Systems, Man, and Cybernetics-Part C: Applications and Reviews, 42*(1), 86–100.
31. Triguero, I., García-Gil, D., Maillo, J., Luengo, J., García, S., & Herrera, F. (2019). Transforming big data into smart data: An insight on the use of the k-nearest neighbors algorithm to obtain quality data. *Wiley Interdisciplinary Reviews: Data Mining and Knowledge Discovery, 9*(2), e1289.
32. Triguero, I., Peralta, D., Bacardit, J., García, S., & Herrera, F. (2015). MRPR: A MapReduce solution for prototype reduction in big data classification. *Neurocomputing, 150*, 331–345.
33. Verbaeten, S., & Assche, A. V. (2003). Ensemble methods for noise elimination in classification problems. In *4th International Workshop on Multiple Classifier Systems*. Lecture Notes on Computer Science (Vol. 2709, pp. 317–325). Berlin: Springer.
34. Wilson, D. L. (1972). Asymptotic properties of nearest neighbor rules using edited data. *IEEE Transactions on Systems, Man, and Cybernetics, 2*(3), 408–421.
35. Wu, X. (1996). *Knowledge acquisition from databases*. Norwood: Ablex Publishing.
36. Zhong, S., Khoshgoftaar, T. M., & Seliya, N. (2004). Analyzing software measurement data with clustering techniques. *IEEE Intelligent Systems, 19*(2), 20–27.
37. Zhu, X., & Wu, X. (2004). Class noise vs. attribute noise: A quantitative study. *Artificial Intelligence Review, 22*, 177–210.

Chapter 7
Big Data Discretization

7.1 Introduction

Data is present in diverse formats, for example, in categorical, numerical, or continuous values. Categorical or nominal values are unsorted, whereas numerical or continuous values are assumed to be sorted or represent ordinal data. It is well known that data mining (DM) algorithms depend very much on the domain and type of data. In this way, the techniques belonging to the field of statistical learning prefer numerical data (i.e., support vector machines and instance-based learning), whereas symbolic learning methods require inherent finite values and also prefer to perform a branch of values that are not ordered (such as in the case of decision trees or rule induction learning). These techniques are either expected to work on discretized data or to be integrated with internal mechanisms to perform discretization.

The process of discretization has aroused general interest in recent years [12, 14] and has become one of the most effective data preprocessing techniques in DM [11]. Roughly speaking, discretization translates quantitative data into qualitative data, procuring a non-overlapping division of a continuous domain. It also ensures an association between each numerical value and a certain interval. Actually, discretization is considered a data reduction mechanism since it diminishes data from a large domain of numeric values to a subset of categorical values.

As mentioned before, there is an obvious need for discrete values in DM as many algorithms explicitly require them to operate. For instance, three of the ten methods pointed out as the top ten in DM [27] demand data discretization in one form or another: C4.5 [18], Apriori [1], and Naïve Bayes [30]. Among its main benefits, discretization causes in learning methods remarkable improvements in learning speed and accuracy. Besides, some decision tree-based algorithms produce shorter, more compact, and accurate results when using discrete values [13, 14].

The ever-growing generation of data on the Internet is leading us to managing huge collections using data analytics solutions. A large proportion of these datasets are formed by continuous features that may require some discretization processing

© Springer Nature Switzerland AG 2020
J. Luengo et al., *Big Data Preprocessing*,
https://doi.org/10.1007/978-3-030-39105-8_7

before the learning phase is accomplished. For instance, real values are present in all the top-5 largest classification datasets from the popular UCI repository [7]. Most of these collections are generated by some sensor-based generation processes, such as those integrated in recent physics experiments (HIGGS, HEPMASS, SUSY) or even in common smartphones (Heterogeneity Activity Recognition dataset). Additionally, current implementations of tree-based and Bayesian learning in MLlib [15] do not natively include discretization, or it is performed in an unsupervised manner (less accurate).

Classical methods have shown not to scale well when dealing with huge data [28]. As the exponential growth of databases started to entail a problem for experts, several parallel discretization methods were appearing in the literature in the last decades, from multi-processor algorithms [6, 16, 29, 32] to GPU-based solutions [5]. However, all of them are doomed to failure whenever the resources from a single machine are surpassed by the magnitude of today's problems. Novel distributed solutions capable of scaling by attaching extra nodes are expected in this new era of Big Data.

Despite this clear need for discretization solutions in Big Data, few distributed approaches have been proposed so far [22]. Researchers and practitioners are then devoted to devise new distributed methods that allow rapid and scalable discretization procedures, that at the same time provide precise discrete representations.

In this chapter, we shall begin by analyzing the early proposals on parallel discretization (Sect. 7.2). We shall continue with a novel distributed design for the Fayyad and Irani discretizer which provides equal discretization schemes as those in the original version (Sect. 7.3). Then, we shall present a parallel implementation of the Chi2 discretizer (Sect. 7.4). Afterwards, we shall analyze an approximate version of Fayyad algorithm contained in a MapReduce associative classifier (Sect. 7.5). Next, an evolutionary multivariate discretizer is presented (Sect. 7.6). Then, we provide an insight into discretization in Big Data streams (Sect. 7.7). Finally, we summarize the chapter and draw some conclusions (Sect. 7.8).

7.2 Parallel and Distributed Discretization

As mentioned before, just a couple of proposals can be found in the literature that address the Big Data discretization problem. This scarcity also extends to the parallel environment where few methods have been thought in decades. This fact cannot be argued to be motivated by the low complexity of discretizers since most of them require as a preliminary requirement to sort all continuous values (loglinear complexity). In fact, sorting is just a preparation step in discretization, to which has to be added iterative merging or splitting, evaluation of points, etc. This implies that current complexity of discretization algorithms ranges from loglinear to beyond.

Cerquides et al. [6] designed a discretization method that relies on Mantaras distance to evaluate the suitability of candidate partitions. This method performs several steps in parallel: sorting (logarithmic), evaluation of points (linear), and partitioning, among others. Parthasarathy et al. [16] designed a 2-dimensional parallel discretizer for streaming systems. To deal with data updates, this incremental method proposes to dynamically maintain some statistics that avoid the complete re-execution of the algorithm.

Zhao et al. [32] presented a parallel discretizer based on z-score idea. This algorithm utilizes dynamic range that reflects the significance of probability distributions instead of standard unsupervised discretization methods. In [29], Yulong et al. presented an efficient two-step parallel discretization algorithm based on dynamic clustering. The algorithm first utilizes dynamic clustering to create a discrete decision table, and then it generates the final set of points using cut importance discretization algorithms.

Cano et al. [5] introduced GPUs in the discretization field by proposing a GPU-based parallel algorithm inspired by the CAIM discretizer. This CAIM version is able to parallelize a large number of operations, such as sorting, the own CAIM criterion, or the final discretization process. For instance, the ordering stage relies on a radix sort implementation on GPUs to achieve linear complexity in sorting. Additionally, the proposed method can work in a coordinate manner on multiple GPU devices.

Additionally, in Apache Spark's Packages, we can find a distributed implementation of the popular Equal Width Discretizer [22]. This package applies the equal width discretization to a Resilient Distributed Dataset (RDD) with a specified number of bins. In Fig. 7.1 we can find a Spark Package associated with this algorithm.

7.3 DMDLP: Distributed Minimum Description Length Principle Discretizer

In this section, we will describe the DMDLP algorithm (Distributed Minimum Description Length Principle), focused on discretization in distributed environments [22]. It is based on the famous Minimum Description Length Principle

Fig. 7.1 Spark package: equal width discretizer

SparkPackages `spark-MDLP-discretization`

This method implements Fayyad's discretizer based on Minimum Description Length Principle (MDLP) in order to treat non discrete datasets from a distributed perspective. It supports sparse data, parallel-processing of attributes, etc.

```
spark-shell --packages sramirez:spark-MDLP- discretization:1.4.1
```

https://spark-packages.org/package/sramirez/spark-MDLP-discretization

Fig. 7.2 Spark package: DMDLP discretizer

(MDLP) information entropy minimization, presented in [10]. In this work, the authors proved that multi-interval extraction of points and the utilization of boundary points can improve subsequent discretization, both in terms of efficiency and error rate.

In this section we present a new exact design for this well-known algorithm for distributed environments, proving its discretization capabilities on today's large problems. This entropy minimization discretizer, called DMDLP, is implemented under the Apache Spark framework. According to the authors, DMDLP is able to perform 270 times faster than the sequential version, and to provide substantial accuracy improvement on raw continuous schemes. In Fig. 7.2 we can find a Spark Package associated with this research in the third-party Apache Spark Repository.

One important point in this adaption is how the original complexity has been distributed. The overall complexity of DMDLP is mainly determined by two time-consuming operations, namely the sorting operation (logarithmic) and the evaluation process (quadratic). For the latter phase, the worst case implies the complete evaluation of entropy for all points.

The sorting operation relies on a complex primitive of Spark, called sortByKey. This samples the dataset and estimates a set of roughly equal ranges for new partitions. Then, a shuffling operation is started to re-distribute the points according to the previous bounds. Once data are re-distributed, a local sorting operation is launched in each partition. The overall complexity for this operation is loglinear.

The evaluation process is mainly formed by two sub-phases: one that groups the tuples by feature, and a map operation that sequentially evaluates the candidate points. Notice that shuffling in the grouping operation is much more lightweight thanks to the previous sorted partitioning. In the map operation, each feature starts an independent process that, that as in the sequential version, is quadratic but divided between all the threads.

Algorithm 1 Main discretization procedure

Input: S Data set	12: $first \leftarrow first\_by\_part(sorted)$
Input: M Feature indexes to discretize	13: $bds \leftarrow get\_boundary(sorted, first)$
Input: mb Maximum number of cut points to	14: $bds \leftarrow$
select	15: **map** $b \in bds$
Input: mc Maximum number of candidates	16: $< (att, point), q > \leftarrow b$
per partition	17: $EMIT < (att, (point, q)) >$
Output: Cut points by feature	18: **end map**
1: $comb \leftarrow$	19: $(SM, BI) \leftarrow divide\_atts(bds, mc)$
2: **map** $s \in S$	20: $sth \leftarrow$
3: $v \leftarrow zeros(|c|)$	21: **map** $sa \in SM$
4: $ci \leftarrow class\_index(v)$	22: $th \leftarrow select\_ths(SM(sa), mb, mc)$
5: $v(ci) \leftarrow 1$	23: $EMIT < (sa, th) >$
6: **for all** $A \in M$ **do**	24: **end map**
7: $EMIT < (A, A(s)), v >$	25: $bth \leftarrow ()$
8: **end for**	26: **for all** $ba \in BI$ **do**
9: **end map**	27: $bth \leftarrow bth + select\_ths(ba, mb, mc)$
10: $distinct \leftarrow reduce(comb, sum\_vectors)$	28: **end for**
11: $sorted \leftarrow sort\_by\_key(distinct)$	29: **return** $(union(bth, sth))$

Also note that the previous complexity is bounded to a single attribute. To avoid repeating the previous process on all attributes, the current design sorts and evaluates the entire set in a single MapReduce pipeline. Only when the number of boundary points in an attribute is higher than the maximum per partition (extremely rare case), iterative evaluation by feature is necessary.

7.3.1   Main Discretization Procedure

Algorithm 1 shows the main procedure in DMDLP. The algorithm below computes the minimum-entropy cut points by feature according to the MDLP criterion. It uses a parameter to limit the maximum number of points to generate for each feature.

The first step creates combinations for each distinct feature value in the original dataset. This MapReduce phase generates tuples with each feature value as key and a histogram counter as value ($< (A, A(s)), v >$). Afterwards, the tuples are aggregated through a function that sums up all partial vectors with the same key, obtaining class contributions for each distinct point. The resulting tuples are sorted by key so that the complete list of distinct values ordered by feature index and feature value is obtained. This structure will be used later to evaluate the complete set of points in a single step. The first point by partition is also saved (line 11) for further processing. Once such information is stored, the boundary points selection process can be launched.

Algorithm 2 Function to generate the boundary points *(get_boundary)*

Input: *points* An RDD of tuples ($<$ 12:   **end if**
($att, point$), q $>$), where *att* represents   13: $< (la, lp), lq > \leftarrow < (a, p), q >$
the feature index, *point* the point to   14: $accq \leftarrow accq + q$
consider and q the class counter.   15:   **end for**
Input: *first* A vector with all first elements   16: $index \leftarrow get\_index(part)$
by partition   17:   **if** $index < npartitions(points)$ **then**
Output: An RDD of points.   18: $< (a, p), q > \leftarrow first(index + 1)$
  1: *boundaries* $\leftarrow$ 19:   **if** $a <> la$ **then**
  2: **map partitions** *part* $\in$ *points*   20: $EMIT < (la, lp), accq >$
  3:    $< (la, lp), lq > \leftarrow next(part)$ 21:   **else**
  4:    $accq \leftarrow lq$ 22: $EMIT < (la, (p+lp)/2), accq >$
  5:    **for all** $< (a, p), q > \in part$ **do**   23:   **end if**
  6:     **if** $a <> la$ **then**   24:   **else**
  7:      $EMIT < (la, lp), accq >$ 25: $EMIT < (la, lp), accq >$
  8:      $accq \leftarrow ()$ 26:   **end if**
  9:     **else if** $is\_boundary(q, lq)$ **then**   27: **end map**
10:      $EMIT < (la, (p+lp)/2), accq >$ 28: **return** ($boundaries$)
11:      $accq \leftarrow ()$

7.3.2 Boundary Points Selection

Algorithm 2 (*get_boundary*) describes the function that selects those points in the borders. This MapReduce function launches an independent sequential process on each partition. This process is described as follows: for each instance, it evaluates whether the feature index is distinct from the index in the previous point; if yes, this emits a tuple with the last point as key and the accumulated class counter as value. This means that a new feature has appeared, and that this point defines the upper limit for the current feature (last threshold). If the previous condition is not met, the algorithm checks whether the current point is boundary with respect to the previous point. If so, this emits a tuple with the midpoint between these points as key and the vector as value.

Finally, some evaluations are performed on the last point in the partitions. These points are compared with the first point in the next partition (previously broadcasted) in order to check whether there exists a transition to a new feature—generate a prev-point tuple—or not—generate a midpoint tuple. Notice that the last case may not generate a boundary point but this step is needed to allow correct evaluations in further partitions. All tuples generated are then joined into a new RDD of boundary points, which is returned to the main algorithm as *bds*.

In Algorithm 1 (line 14), the *bds* variable is transformed by using a Map function, transforming the previous key to a new key with the feature index ($< (att, (point, q)) >$). This allows us to group the tuples by feature so that each can be processed independently. The *divide_atts* function aims at dividing the tuples into two groups (*big* and *small*) according to the amount of candidate points by feature (count operation). Features in each group will be filtered and treated

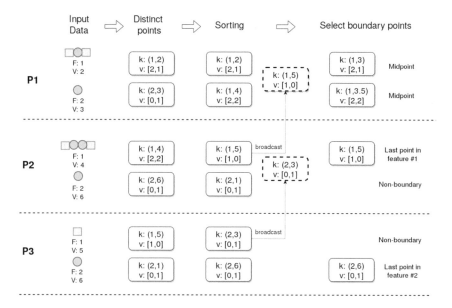

Fig. 7.3 Boundary points generation process. P represents data partitions, k represents keys in pairs, and v values in these pairs

differently according to whether the total number of points for a given feature exceeds a given threshold mc or not. "Small" features will be grouped by key so that these can be processed in a parallel way. The subsequent tuples are now reformatted as follows: $(< point, q >)$. The remaining features are processed in an iterative manner.

The entire boundary points selection process is depicted in Fig. 7.3. In this figure, six distinct points are distributed into three data partitions. The points are sorted by key (feature index and point value) to perform the evaluation. The first points in each partition are sent to the next one to perform the evaluation of limit points. As a result, four boundary points are generated, some are midpoints, and the remaining ones are the upper limits for two features.

7.3.3   DMDLP Evaluation

As mentioned before, each group of features is evaluated in a different manner. Small features are evaluated in a single MapReduce stage where each feature corresponds with a single partition, whereas big features are evaluated iteratively as each feature corresponds with a complete RDD with several partitions. The first option is obviously more efficient, however, the second one is less frequent due to the number of candidate points is typically small according to the experiments

---

Algorithm 3 Recursively evaluation of partitions *(select_ths)*

Input: *cands* A RDD/array of tuples (< *point, q* >), where *point* represents a candidate point to evaluate and *q* the class counter.

Input: *mb* Maximum number of intervals or bins to select

Input: *mc* Maximum number of candidates to eval in a partition

Output: An array of thresholds for a given feature

1: $st \leftarrow enqueue(st, (candidates, ()))$
2: $result \leftarrow ()$
3: **while** $|st| > 0$ & $|result| < mb$ **do**
4:     $(set, lth) \leftarrow dequeue(st)$
5:     **if** $|set| > 0$ **then**

6:       **if** $type(set) = {'array'}$ **then**
7:         $bd \leftarrow arr\_select\_ths(set, lth)$
8:       **else**
9:         $bd \leftarrow rdd\_select\_ths(set, lth, mc)$
10:       **end if**
11:       **if** $bd <> ()$ **then**
12:         $result \leftarrow result + bd$
13:         $(left, right) \leftarrow divide(set, bd)$
14:         $st \leftarrow enqueue(st, (left, bd))$
15:         $st \leftarrow enqueue(st, (right, bd))$
16:       **end if**
17:     **end if**
18: **end while**
19: **return** $(sort(result))$

Algorithm 4 Function that computes statistics for all candidate partition and select the best cut point (sequential version) *(arr_select_ths)*

Input: *cands* An array of tuples (< *point, q* >), where *point* represents a candidate point to evaluate and *q* the class counter.

Output: The minimum-entropy cut point

1: $total \leftarrow sum\_freqs(cands)$
2: $lacc \leftarrow ()$

3: **for** $< point, q > \in cands$ **do**
4:     $lacc \leftarrow lacc + q$
5:     $freqs \leftarrow freqs + (point, q, lacc, total - lacc)$
6: **end for**
7: **return** $(select\_best(cands, freqs))$

---

in [22]. In both cases, the *select_ths* function serves to evaluate and select the most relevant cut points from the candidate set. For small features, a Map function is applied on each partition *(arr_select_ths)*. For big features, the process is more complex, and each feature needs a complete iteration over a different RDD *(rdd_select_ths)*.

Algorithm 3 *(select_ths)* recursively selects the most promising candidates according to the MDLP criterion (single-step version). This algorithm starts by selecting the best cut point in the entire set. If the criterion accepts the addition, the point is removed from the candidate set and included into the final scheme. The candidate set is then divided into two new blocks using this cut point. Both partitions are then evaluated recursively starting from the left block and repeating the previous process. This process finishes when there is no partition to evaluate or the maximum number of selections is met.

Algorithm 4 *(arr_select_ths)* shows the sequential process (small features) in charge of generating sufficient statistics for the selection process. This version is much simpler than the RDD version as it works in a sequential manner. Firstly, it computes the class contributions on the entire set. Afterwards, a new iteration is performed to obtain the left and right histograms for each possible partition. This

---

Algorithm 5 Function that computes statistics for all candidate partition and select the best cut point (RDD version) *(rdd_select_ths)*

Input: *cands* An RDD of tuples (<
   point, q >), where *point* represents a
   candidate point to evaluate and *q* the class
   counter.
Input: *mc* Maximum number of candidates to
   eval in a partition
Output: The minimum-entropy cut point
1: *npart ← round(|cands|/mc)*
2: *cands ← coalesce(cands, npart)*
3: *totalpart ←*
4: **map partitions** *partition ∈ cands*
5:    *return(sum(partition))*
6: **end map**
7: *total ← sum(totalpart)*
8: *freqs ←*

9: **map partitions** *partition ∈ cands*
10:    *index ← get_index(partition)*
11:    *ltotal ← ()*
12:    *freqs ← ()*
13:    **for** *i = 0 until index* **do**
14:       *ltotal ← ltotal + totalpart(i)*
15:    **end for**
16:    **for all** *< point, q >∈ partition* **do**
17:       *freqs ← freqs+(point, q, ltotal+*
          q, total − ltotal)
18:    **end for**
19:    *return(freqs)*
20: **end map**
21: **return** *(select_best(cands, freqs))*

---

is done by aggregating the vectors from the leftmost point to the current point, and from this to the right-most point. Once the accumulated histogram for each candidate point are calculated (in form of $< point, q, lq, rq >$), the algorithm evaluates the candidates using the *select_best* function.

Algorithm 5 (*rdd_select_ths*) shows the selection process for "big" features (points in RDD). This process is performed in a distributed manner since the number of candidate points exceeds the maximum size allowed. For each feature, the set of points is re-distributed to a better partition scheme that homogenizes the amount of elements by partition and node (*coalesce* function, line 1–2). Afterwards, a parallel process is launched to compute the accumulated histogram by partition. The results are then aggregated to obtain the total accumulated frequency for the entire set. In line 9, a distributed process computes the class histogram for each candidate point (Algorithm 4). In this procedure, the program sums up all the contributions from the previous (left) partitions to the current one in order to obtain the starting left accumulator (*ltotal*). The statistics for each inner point are computed using the local contributions and the partial accumulator initialized above (line 7). Once statistics are computed ($< point, q, lq, rq >$), the algorithm evaluates all candidate points in the same manner as in the sequential version.

Algorithm 6 determines if a given point is accepted or not according to the MDL principle. Thus, for each point,[1] the entropy is computed for the two partitions associated with each point (line 8) as well as the total entropy for the entire set (lines 1–2). Using these values, the entropy gain for each point and its MDLP score is obtained. If the point is accepted by MDLP, the algorithm emits a tuple with

---

[1]If the points are in array format, a loop is used to evaluate points, else a distributed map function is used instead.

Algorithm 6 Function that determines the acceptance or not of points according to the MDL principle *(select_best)*

Input: $freqs$ An array/RDD of tuples ($<$ $point, q, lq, rq >$), where point represents a candidate point to evaluate, leftq the left accumulated frequency, rightq the right accumulated frequency and q the class frequency counter.
Input: $total$ Class frequency counter for all the elements
Output: The minimum-entropy cut point
1: $n \leftarrow sum(total)$
2: $totalent \leftarrow ent(total, n)$
3: $k \leftarrow |total|$
4: $accp \leftarrow ()$
5: **for all** $< point, q, lq, rq > \in freqs$ **do**
6: $\quad k1 \leftarrow |lq|; k2 \leftarrow |rq|$
7: $\quad s1 \leftarrow sum(lq); s2 \leftarrow sum(rq);$
8: $\quad ent1 \leftarrow ent(s1, k1); ent2 \leftarrow ent(s2, k2)$
9: $\quad parent \leftarrow (s1 * ent1 + s2 * ent2)/s$
10: $\quad gain \leftarrow totalent - parent$
11: $\quad delta \leftarrow log_2(3^k - 2) - (k * hs - k1 * ent1 - k2 * ent2)$
12: $\quad accepted \leftarrow gain > ((log_2(s - 1)) + delta)/n$
13: $\quad$ **if** $accepted = true$ **then**
14: $\quad\quad accp \leftarrow accp + (parent, point)$
15: $\quad$ **end if**
16: **end for**
17: **return** $(min(accp))$

the weighted entropy average for the resulting partitions and the point itself. From the set of accepted points, the algorithm selects the one with the minimum class information entropy.

The results produced by both groups (small and big) are finally joined into the discretization scheme.

7.4 DChi2: Distributed Chi2 Discretizer

In this section, we will describe the DChi2 algorithm (Distributed Chi2 Discretizer), focused on a MapReduce implementation of Chi2 algorithm [31]. Chi2 algorithm is a global supervised bottom-up discretizer based on $\chi2$ statistical test. This can be seen as an automatic version of ChiMerge where the statistical significance level is adjusted to merge more and more adjacent intervals as long as the inconsistency criterion is satisfied. Here, the stopping criterion is achieved when there are enough inconsistencies present in data considering a limit of zero or δ level as default.

Chi2 consists of two main phases, in which computations are performed on each input attribute, and therefore can be easily parallelized. The first phase can be seen as a generalization of ChiMerge, whereas the second phase performs small tunes on the intervals. Namely, the first phase works on a global significance level, whereas the second phase works on separated significance levels for each attribute.

The idea proposed by Zhang was to organize Chi2 into two MapReduce phases: one stage that reads raw data, and computes distinct points and class contributions; and another stage that applies the Chi2 algorithm on the entire set of distinct points. Figure 7.4 depicts the scheme proposed by Zhang in a MapReduce logical format.

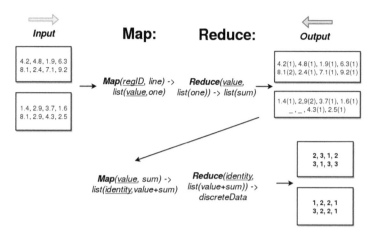

Fig. 7.4 Distributed Chi2 discretizer. It consists of two MapReduce phases: one that computes distinct points in a parallel manner, and another that applies Chi2 on the entire set of points

As can be noticed from the previous scheme, only the complexity burden associate to the generation of distinct points is distributed across the cluster. No single part of Chi2 is accelerated using a distributed scheme. This implementation can thus be deemed as a preliminary approach in which further improvements should be concentrated in the parallelization of Chi2.

7.5 Distributed Discretization on MapReduce Associative Classification

In this section, we will describe the MRAC algorithm (MapReduce Associate Classifier), an association rule-based classifier built upon MapReduce [4]. The proposed method infers classification associative rules using a distributed implementation of the well-known FP-Growth algorithm. Once the corresponding rules have been generated, a distributed pruning process is performed. The resulting set of elected rules is finally utilized to predict unlabeled elements.

As part of MRAC, the authors designed an approximation of the MDLP algorithm to transform continuous values into discrete ranges, which is required by the rule-based inference process. This discretization process is composed of two MapReduce phases, as shown in Fig. 7.5.

In the first MapReduce stage, each Mapper reads different groups of examples from HDFS according to a percentage defined as input parameter, and applies equifrequency discretization on each group. The subsequent output is the sorted list of cut points for each Mapper. The Reducers fused all the lists into a final sorted list of equifrequency bin boundaries.

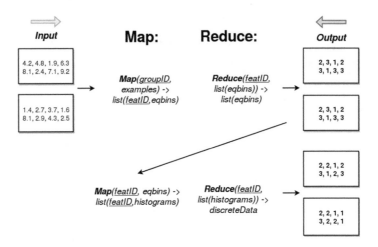

Fig. 7.5 Distributed discretization in MRAC, composed of two MapReduce phases: one that generates equifrequency boundaries, and another that computes class histograms based upon these bins and selects the best points according to the MDLP principle

In the second stage, the Mappers compute class histograms for all the bins generated by fetching again the entire dataset. Then each Mapper calculates the percentage of instances belonging to each class. Finally, the Reducers aggregate the histograms and select the most relevant cut points following to the MDLP principle.

The approximate version presented above reduces the number of cut points to consider thanks to the former unsupervised discretization step. This makes discretization in MRAC more rapid than in DMDLP (Sect. 7.3), but much less accurate as it utilizes a decimated subset of candidate points. Furthermore, the boundary optimization as described in [10] is not guaranteed by MRAC since initial points are generated in an unsupervised manner without considering class borders. In MRAC, the higher the frequency used in the bins, the more accurate the approximation will be.

Finally, MRAC requires to replicate the entire set of points in order to count class contributions before MDLP is applied. In case of complex borders with a massive amount of boundaries, MRAC would perform poorly due to high network cost of replication.

7.6 A Distributed Evolutionary Multivariate Discretizer for Big Data

In this section, we will describe the DEMD algorithm (Distributed Evolutionary Multivariate Discretizer), a discretizer based on an evolutionary points selection scheme [20]. It is based on the classical EMD discretizer [19]. EMD is an

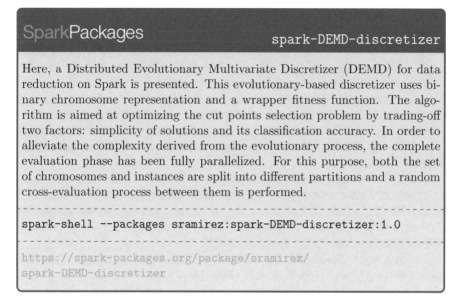

Fig. 7.6 Spark package: DEMD discretizer

evolutionary-based discretizer with binary representation and a wrapper fitness function. Although both algorithms share some common aspects (like representation and fitness function), DEMD goes beyond a simple parallelization, and offers an approximative, scalable, and resilient solution to deal with the Big Data discretization problem. Alike EMD, in DEMD, partial solutions are generated locally, and eventually fused to produce the final discretization scheme. This proposal is the first evolutionary approach in dealing with the large-scale discretization problem. In Fig. 7.6 we can find a Spark Package associated with this research in the third-party Apache Spark Repository.

This section analyzes the design of the distributed discretization solution for Big Data. This algorithm splits both the set of cut points and instances into partitions, and evaluates them through a cross-evaluation system. With this distributed scheme we maximize the resource usage throughout the entire process. If a point is selected by one of the evaluation processes, it counts as a single vote. All these votes are aggregated to obtain the final score per point. Finally, the final discretization scheme is obtained through a voting scheme.

In Sect. 7.6.1, the main procedure in charge of partitioning the instances are feature, and aggregating the partial solutions is presented. Section 7.6.2 illustrates the process of computing boundary points. Section 7.6.3 exposes how the chromosomes are evaluated in a distributed manner by using EMD.

7.6.1 Main Discretization Procedure

Procedure 7 shows the main procedure of the discretization algorithm. Hereafter we will use the term partition to describe the data partitions, and the term chunk to describe the feature partitions. This procedure is in charge of distributing the initial cut points (computed in Sect. 7.6.2) among the set of chunks. The partitions already created are associated with these chunks so that each chunk is evaluated on the instances contained in one or more partitions. After the parallel selection process is performed (in Sect. 7.6.3), this procedure creates the final matrix of selected cut points.

The first step computes the boundary points (BF) in a distributed way using the function $getBoundary$ (line 1, Sect. 7.6.2). Each tuple in BF consists of a feature ID (fid) and a list of points. Based on this variable, DEMD creates FI (feature information), and BP (boundary points per feature). All this information will serve us to create the chromosome chunks.

The procedure divides the evaluation of cut points using subsets of features (called chunks) (lines 2–13). To do that, DEMD first sorts all features by the number of boundary points contained in each one (ascending order). Then, DEMD computes the number of chunks (ncp) in which the entire list of boundary points will be divided. ncp is computed using several variables which are related according to Eq. (7.1).

Algorithm 7 Main discretization procedure

Input: D dataset	16: **for all** $w \in windows$ **do**
Input: M Feature indexes to discretize	17: $p \leftarrow shuffle(w)$
Input: uf Multivariate user factor	18: **for** $i = 0 \rightarrow i < p.size$ **do**
Input: alp Alpha parameter	19: $CH(i).add(p(i))$
Input: ne Number of evaluations	20: **end for**
Input: sr Sampling rate	21: **end for**
Input: vp Percentage of selected points	22: $CH \leftarrow broadcast(CH)$
Output: Cut points by feature	23: $SD \leftarrow stratifiedSampling(D, sr)$
1: $BF \leftarrow getBoundary(D, M)$	24: $SP \leftarrow select(SD, CH, uf, alp, sr, vp)$
2: $BP \leftarrow ()$; $FI \leftarrow ()$;	25: $TH \leftarrow ()$
3: **for all** $< fid, l > \in BF$ **do**	26: **for** $< chid, lf > \in SP$ **do**
4: $BP(fid) = l$	27: $ind \leftarrow 0$; $chunk \leftarrow CH(chid)$
5: $FI.add(Feature(fid, l.size))$	28: **for** $feat \in chunk$ **do**
6: **end for**	29: **for** $i = 0 \rightarrow i < feat.size$ **do**
7: $FI \leftarrow broadcast(sortBySize(FI))$	30: **if** $lf(i + ind) == true$ **then**
8: $BP \leftarrow broadcast(BP)$	31: $point \leftarrow BP(feat.id)(i + ind)$
9: $nbp \leftarrow totalSize(BP)$	32: $TH(feat.id).add(point)$
10: $ds \leftarrow nbp/D.npartitions$	33: **end if**
11: $ms \leftarrow max(FI(0).size, ds)$	34: **end for**
12: $df \leftarrow max(uf, ms/ds)$	35: $ind \leftarrow ind + feat.size$
13: $ncp \leftarrow nbpoints/(df * ds)$	36: **end for**
14: $windows \leftarrow makeGroups(FI, ncp)$	37: **end for**
15: $CH \leftarrow ()$	38: **return** (TH)

$$ncp = np/(max(uf, ms/ds) \cdot ds), \qquad (7.1)$$

where np is the total number of boundary points, ds the current proportion of points by data partition, uf the split factor specified by the user, and ms the maximum between the largest feature size and ds.

Usually each feature is contained in a single chunk, but it may change in case the user specifies a greater value, or the largest feature surpasses the default size since points belonging to the same feature cannot be separated. In the latter case, a finer-grained division will be performed, which means more chunks. This scenario normally entails a quicker evaluation, but a loss in effectiveness.

The evaluation procedure starts to distribute points between the chunks (CH) (lines 14–21). In each iteration, a group of nc features is collected and randomly distributed among the chunks. The loop ends when there is no feature to collect. This mechanism will enable a fairly distribution of boundary points, without points from the same feature in different chunks, and with a similar number of features per chunk.

Once the distribution of points is completed, a stratified sampling process (by class) is performed on D (line 23). The resulting sample SD is used to evaluate the boundary points in a distributed manner. According to the multivariate factor $(max(uf, ms/ds))$, each partition randomly selects as many chunks as indicated by these factor (usually only one). Then, each partition is responsible of evaluating the points contained in their associated chunks (line 24). The selection phase is described in detail in Sect. 7.6.3.

Each selection process returns its aggregated partial solution (the best chromosome per chunk), and saves the tuples (chunk ID, best solution) in SP. All these partial results are then summarized using a voting scheme, considering the threshold (vp). Finally, the main procedure processes the binary vectors to obtain the final matrix of cut points (TH) (line 26–38). This procedure fetches the features in each chunk, and its correspondent points. If a given point has been selected, it is added to the final matrix. If not, this is omitted.

An illustrative scheme of the entire process is detailed in Fig. 7.7. In this example, there are four features with different amounts of boundary points (8, 5, 4, 10). Boundary points are then uniformly distributed into three chunks where features may be mixed, like in chunk $C1$. Afterwards chromosome chunks are grouped with seven data partitions following a correspondence table that relates chunks and partitions according to the multivariate factor. Once local evaluation threads have ended, partial discretization results (binary vectors) for the same chromosome part are aggregated by summing votes. Most-voted points in each chunk according to vp (proportion of points to select) are selected, and adapted to create the global selection matrix.

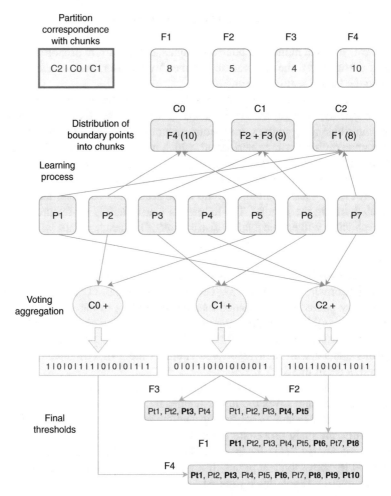

Fig. 7.7 A simplified representation of the DEMD process. F represent the features, C the chromosome chunks, P the dataset partitions to evaluate, and Pt the boundary points. The selected points have been highlighted in bold

7.6.2 Computing the Boundary Points

Procedure 8 (*getBoundary*) describes the function that computes border points in data. This procedure consists of three steps. Firstly, the distinct points (D) in the dataset are calculated by removing duplicated elements. Secondly, the resulting points are sorted (S) and distributed by feature index so that all the points from the same feature will not be separated. Finally, the boundary points (BP) in each feature are evaluated sequentially.

Algorithm 8 Function to generate the boundary points *(getBoundary)*

Input: D dataset
Input: M Feature indexes to discretize
Output: The set of boundary points (feature index, point value).

1: $CB \leftarrow$
2: **map** $s \in D$
3:    $v \leftarrow zeros(|c|)$
4:    $ci \leftarrow classIndex(v)$
5:    $v(ci) \leftarrow 1$
6:    **for all** $A \in M$ **do**
7:       $EMIT < (A, A(s)), v >$
8:    **end for**
9: **end map**
10: $D \leftarrow reduce(CB, sumVectors)$
11: $S \leftarrow sortByKey(D)$
12: $FP \leftarrow firstByPart(S)$
13: $BP \leftarrow$
14: **map partitions** $PT \in S$
15:    $< (la, lp), lq > \leftarrow next(PT)$
16:    **for all** $< (a, p), q > \in PT$ **do**

17:       **if** $a <> la$ **then**
18:          $EMIT < la, lp >$
19:       **else if** $isBoundary(q, lq)$ **then**
20:          $EMIT < la, (p + lp)/2 >$
21:       **end if**
22:       $< la, lp > \leftarrow < a, p >$
23:    **end for**
24:    $index \leftarrow getIndex(PT)$
25:    **if** $index < npartitions(S)$ **then**
26:       $< (a, p), q > \leftarrow FP(index + 1)$
27:       **if** $a <> la$ **then**
28:          $EMIT < la, lp >$
29:       **else**
30:          $EMIT < la, (p + lp)/2 >$
31:       **end if**
32:    **else**
33:       $EMIT < la, lp >$
34:    **end if**
35: **end map**
36: **return** $(BP.groupByKey())$

The procedure starts by launching a parallel process on each partition (taking advantage of data locality) with the aim of computing the distinct points (CB) (lines 1–9). Once the points are sorted (D) and the first point by partition is distributed (FP), DEMD evaluates whether each points belong to any border as follows (lines 13–36): for each point, it checks whether the feature index is distinct from the index of the previous point; if it is so, DEMD generates a tuple with the feature index of the last point as key, and its correspondent value as value. By doing so, the last point from the current feature is always kept as the last threshold. If there are more points in this feature, the procedure evaluates whether the current point accomplishes the boundary condition with respect to the previous point. If it is so, this generates a tuple with the feature index as key, and the midpoint between these two points as value.

The last point in each partition is considered as a special case (lines 25–34). These points are compared with the first point in the following partition (broadcasted). If the feature indexes are different, the procedure emits a tuple with the last point. If not and the point is boundary, DEMD emits a tuple with the midpoint between these two points. Finally, all the tuples generated in each partition are joined into a RDD of boundary points, which is returned to the main procedure.

The previous process is depicted in Fig. 7.8. In this figure we can see three partitions with six different points. The points are sorted by key (feature index and point value) to perform the evaluation. The first point for each partition is sent to the following partition to perform the evaluation of the last points. As a result, three boundary points are generated, some are midpoints and some are the last points in features.

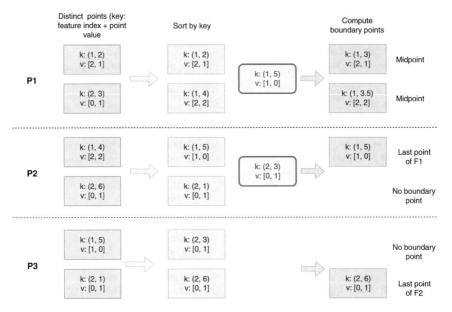

Fig. 7.8 Distributed computation of boundary points. P represents the partitions. The points broadcasted have been highlighted in bold

7.6.3 Distributed Cut Points Selection

Procedure 9 shows the distributed operations used to aggregate the solutions generated by each local evaluation process, and to decide the final discretization scheme. Note that local evaluation of points is performed by launching a single instance of EMD on each data partition. This process consists of two steps: the first one starts a selection process (map) on each pair chunk-partition and aggregates the subsequent solutions to produce the final number of votes. The second step is aimed at selecting the most-voted points by chunk according to the threshold (vp) defined by the user.

Firstly, each chunk is associated with one or more data partitions using PO, which is a table formed by tuples (chunk ID, data partition). For each tuple, a map operation is started (lines 3–22). This map operation starts by creating a data matrix with the instances contained on each partition and those features present in the chunk. Afterwards, the procedure executes an evaluation thread on each submatrix (FD) in order to evaluate the corresponding boundary points (CP). As a result, the best chromosome (a binary vector) in the population is returned (BI).

The binary vector will be transformed into a numeric vector to annotate number of selections (CO) (lines 13–22). The final result emitted by the partition is a tuple with the identifier of the chunk as key, and the vector count—number of times each point has been selected—and a chunk count—maximum number of partitions in

---

Algorithm 9 Function to perform the evolutionary selection process *(select)*

Input: *SM* Sampled boundary points	12: $BI \leftarrow EMD(FD, CP, alpha, ne)$
Input: *CH* Feature chunks	13: $CO \leftarrow ()$
Input: *uf* Multivariate user factor	14: **for** $i = 0 \rightarrow i < BI.size$ **do**
Input: *alpha* Alpha parameter (evolutionary	15: **if** $BI(i) == 1$ **then**
process)	16: $CO(i) = 1$
Input: *ne* Number of evaluations (evolution-	17: **else**
ary process)	18: $CO(i) = 0$
Input: *sr* Sampling rate	19: **end if**
Input: *vp* Percentage of selected points	20: **end for**
Output: Cut points by feature	21: $EMIT < chid, (CO, 1) >$
1: $PO \leftarrow shuffle(seq(0, CH.size))$	22: **end map**
2: $R \leftarrow$	23: $CO \leftarrow R.reduceByKey(sum())$
3: **map partitions** $< index, DT > \in SM$	24: $SL \leftarrow$
4: $chid \leftarrow PO(index \% CH.size)$	25: **map** $< chid, (AC, c) > \in CO$
5: $C \leftarrow CH(chid)$	26: $S \leftarrow sort(AC)$
6: $npoints \leftarrow totalSize(chunk)$	27: $ps \leftarrow AC.size * vp$
7: $CP \leftarrow ()$; $FD \leftarrow ()$	28: $BA \leftarrow take(S, ps)$
8: **for** $i = 0 \rightarrow i < chunk.size$ **do**	29: $EMIT < chid, BA >$
9: $FD(i) \leftarrow DT(i)$	30: **end map**
10: $CP(i) \leftarrow C(i)$	31: **return** (SL)
11: **end for**	

---

which has been evaluated—as value. This procedure will indicate the selection ratio
for each point. The partial values generated above are aggregated by reducing the
tuples by key.

Secondly, the procedure starts a map operation (lines 25–30) to select the most-
voted points by chunk (SL). The procedure orders all the points by number of votes,
and selects in order as much points as specified by vp. The result is a tuple with
chunk ID as key, and the selection vector as value. Finally, previous results will
eventually be transformed to a matrix of points in the main procedure.

7.7 Discretization in Big Data Streaming

Discretization is one of the most extended data preprocessing techniques. Although
we can find many proposals for static Big Data preprocessing, there is little research
devoted to the continuous Big Data problem. Apache Flink is a recent and novel
Big Data framework, following the MapReduce paradigm, focused on distributed
stream and batch data processing [3].

In this section, we will describe a data stream library named DPASF (Data
Preprocessing Algorithms for Streaming in Flink), focused on Big Data stream
preprocessing [2]. The library is composed of six of the most popular and widely
used data preprocessing algorithms for Apache Flink. It contains three algorithms
for performing feature selection, and three algorithms for discretization. In this

FlinkML   DPASF

Big Data library oriented to online data preprocessing for Apache Flink. This
library contains six of the most popular and widely used algorithms for data
preprocessing in data streaming. It is composed of three feature selection
algorithms and three discretization algorithms.

https://sci2s.ugr.es/BigDaPFlink

Fig. 7.9 FlinkML: DPASF library

section, we focus on the three online discretization algorithms implemented in the
library. In Fig. 7.9 we can find the algorithm implementations associated with this
research.

7.7.1   IDA: Incremental Discretization Algorithm

IDA [25] uses a random sample of the data stream for performing an approximate
quantile-based discretization. This random sample is used to calculate the cut points.

IDA uses a random sample of the data because it is not feasible nor possible
for high-throughput data streams to maintain a complete record of all the values
processed. The sample method used is called reservoir sampling [24]. It maintains a
random sample of s values V_i for each attribute X_i. The first s values that arrive for
each X_i are added to its corresponding V_i. Thereafter, every time a new instance
$\langle \mathbf{x}_n, y_n \rangle$ arrives, each of its values $\mathbf{x}_n^i$ replace a randomly selected value of the
corresponding V_i with probability s/n.

Each value of each attribute is stored in a vector of interval heaps [23]. V_i^j stores
the values for the jth bin of X_i. The reason to use an Interval Heap is that it provides
efficient access to minimum and maximum values in the heap and direct access to
random elements within the heap.

Algorithm 10 shows the pseudocode for IDA. This algorithm first computes the
cut points for the dataset with the desired number of bins. In order to compute
them, each instance is mapped to the result of IDA, which returns the computed
cut points. To do so, each feature is zipped with its index, and then folded with its
corresponding class label and a zero feature vector that will be filled in each iteration
of the fold operation, with the returned value of the IDA algorithm. Once cut points
are stored, line 6 in Algorithm 10 discretizes the data according to those cut points.

Algorithm 10 IDA Algorithm

Input: *data* a DataSet LabeledVector (label, features)
Input: *bins* number of bins
Output: Discretized dataset with desired number of *bins*
 1: *cuts* ←
 2: **map** $((y, x) \in data)$
 3:    *zipped* ← $zipWithIndex(x)$
 4:    $FoldLeft((y, emptyfeature), IDA())$
 5: **end map**
 6: **return** $discretize(data, cuts)$

7.7.2 PiD: Partition Incremental Discretization Algorithm

PiD [17] is a discretization algorithm that performs an incremental discretization. This discretization is performed in two tasks or layers. The first layer receives the sequence of input data and the range of the variable and keeps some statistics on the data using many more intervals than required. The range of the variable is used to initialize the cut points with the same width. Each time a new value arrives, this layer is updated in order to compute the corresponding interval for the value. Each interval has an internal count of the values it has seen so far. When a counter for an interval reach a threshold, a split process is triggered to generate two new intervals. If the interval triggering the split process is the last or the first, a new interval with the same step is created. Otherwise the interval is split in two. In summary, the first layer simplifies and summarizes the data.

The second layer creates the final discretization based on the statistics calculated by the first layer of the architecture. This architecture processes streaming examples in a single scan, in constant time and space even for infinite sequences of examples. To do so, the second layer merges the set of intervals computed in the previous layer.

PiD stores the information about the number of examples per class in each interval in a matrix. In this matrix, columns correspond with the number of intervals and rows with the number of classes. With this information, the conditional probability of an attribute belonging to an interval given that the corresponding example belongs to a class can be computed as $P(b_i < x \le b_{i+1}|Class_j)$.

To perform the actual discretization, PiD uses *Recursive entropy discretization* [9]. This algorithm was developed by Fayyad and Irani [8]. It uses the class information entropy of two candidate partitions to select the boundaries for discretization. It starts searching for a unique threshold that minimizes the entropy over all possible thresholds, then, it is applied recursively to both partitions. It uses the *minimum description length* [26] principle as stop criteria. The algorithm works as follows:

First, the entropy before and after the split is computed as well as its information gain. Then, the entropy for both left and right splits is computed. Finally the algorithm checks if the split is accepted using the following formula:

Algorithm 11 PiD Algorithm

Input: *data* a DataSet LabeledVector (label, features)
Input: α parameter
Input: *step* parameter
Output: Discretized dataset
 1: *cuts* $\leftarrow$
 2: **map** *instance* $\in$ *data*
 3:    *(instance, Histogram*, 1)
 4: **end map**
 5: **reduce** $(m1, m2) \in cuts$
 6:    $UpdateLayer1(m1, m2)$
 7:    $UpdateLayer2(m1, m2)$
 8: **end reduce**
 9: **return** *discretize(data, cuts)*

$$Gain(A, T; S) < \frac{\log_2(N-1)}{N} + \frac{\Delta(A, T; S)}{N}, \tag{7.2}$$

where N is the number of instances in the set S,

$$Gain(A, T; S) = H(S) - H(A, T; S) \tag{7.3}$$

and

$$\Delta(A, T; S) = \log_2(3^k - 2) - [k \cdot H(S) - k_1 \cdot H(S_1) - k_2 \cdot H(S_2)], \tag{7.4}$$

where k_i is the number of class labels represented in the set S_i.

Algorithm 11 shows the pseudocode for PiD, this algorithm first initializes the required data structures using a map function. This map function expands the dataset and adds to it a histogram and a count of the total number of instances seen so far. Then this data is reduced, computing in each reduce step the layers one and two as described in the original algorithm. Once this reduce stage has been completed, it returns the discretized data using the previously computed cut points.

7.7.3 LOFD: Local Online Fusion Discretizer

LOFD [21] is a novel, online, and self-adaptive discretization solution for streaming classification which aims at reducing the negative impact of fluctuations in evolving intervals. LOFD mainly relies on highly-informative class statistics to generate accurate intervals at every step. Furthermore, local nature of operations implemented in LOFD offers low response times, thereby making it suitable for high-speed streaming systems.

The algorithm is composed of two phases, the main process, at instance level, and the merge/split process, at interval level. The main process works as follows. First, discrete interval is initialized following the static process defined in [33]. The discretization is then performed on the first *initTh* instances. From that moment on, LOFD updates the scheme of intervals in each iteration and for each attribute. For each new instance, it retrieves its ceiling interval (implemented as a red-black tree). If the point is above the upper limit a new interval is generated at that point, making that point the new maximum for the current attribute. A merge between the old and the new last interval is evaluated by computing the quadratic entropy, if the result is lower than the sum of both parts, the merge is accepted.

Finally, each point is added to a queue with a timestamp to control future removals in case the histogram overflows. If necessary, LOFD recovers points from the queue in ascending order and removes them until there is space left in the histogram.

The split/merge phase is triggered each time a boundary point is processed. The new boundary point splits an interval in two, one interval contains the points in the histogram with values less than or equal to the new point and keeps the same label. Each time a new interval is generated, the merge process is triggered for the intervals being divided and their neighbors.

Algorithm 12 shows pseudocode for LOFD. This algorithm first instantiates a LOFD helper, and maps the data according to the computed cut points this helper returns. Once all cut points have been collected, the reduce function extracts only the most recently computed cut points. Finally, it performs the discretization based on those same cut points.

Algorithm 12 LOFD Algorithm

Input: *data* a DataSet LabeledVector (label, features)
Output: Discretized dataset
1: $lofd \leftarrow LOFDInstance$
2: $cuts \leftarrow$
3: **map** $x \in data$
4:     $discretized \leftarrow lofd.applyDiscretization(x)$
5:     **for** $s = 0$ *until discretized.size* **do**
6:         $lofd.getCutpoints(s)$
7:     **end for**
8: **end map**
9: **reduce** $(\_, b) \in cuts$
10:     b
11: **end reduce**
12: **return** $discretize(data, cuts)$

7.8 Summary and Conclusions

Discretization is one of the most studied techniques in data preprocessing. It is focused on transforming the values of the instances from continuous values to discrete ones. This reduces the complexity of the data and also removes outliers.

This chapter presents the proposals devoted to perform discretization in Big Data scenarios. A distributed version of one of the most popular discretizers (MDLP) is presented. A distributed implementation of Chi2 algorithm and an approach for performing discretization with association rules are also introduced. Next, we describe one of the most recent proposals for distributed discretization, DEMD. Finally, we analyze the proposals for dealing with discretization in Big Data streaming. These techniques have the challenge of having to adapt to the evolving data and concept drift.

Discretization is one of the most complex tasks regarding data preprocessing. The wrong cut points selection will lead to the inability to learn or to acquire valuable knowledge from the data. For this reason, data discretization taxonomies have been created in order to select the best method for a given problem [22]. However, the lack of Big Data discretizers is still an open issue, due to the complexity of finding suitable cut points. While normal-sized data shows a rich ecosystem of discretized, based on many different criteria, more algorithms are needed for tackling the Big Data discretization problem from such a diverse taxonomy.

References

1. Agrawal, R., & Srikant, R. (1994). Fast algorithms for mining association rules. In *Proceedings of the 20th Very Large Data Bases Conference (VLDB)* (pp. 487–499).
2. Alcalde-Barros, A., García-Gil, D., García, S., & Herrera, F. (2019). DPASF: A Flink library for streaming data preprocessing. *Big Data Analytics, 4*(1), 4.
3. Apache Flink. (2019). Apache Flink. http://flink.apache.org/.
4. Bechini, A., Marcelloni, F., & Segatori, A. (2016). A MapReduce solution for associative classification of big data. *Information Sciences, 332*, 33–55.
5. Cano, A., Ventura, S., & Cios, K. J. (2014). Scalable CAIM discretization on multiple GPUs using concurrent kernels. *The Journal of Supercomputing, 69*(1), 273–292.
6. Cerquides, J., & de Mántaras, R. L. (1997). Proposal and empirical comparison of a parallelizable distance-based discretization method. In *Proceedings of the Third International Conference on Knowledge Discovery and Data Mining, KDD'97* (pp. 139–142).
7. Dua, D., & Graff, C. (2017). UCI machine learning repository. http://archive.ics.uci.edu/ml.
8. Fayyad, U. M., & Irani, K. B. (1993). Multi-interval discretization of continuous-valued attributes for classification learning. In *IJCAI*.
9. Fayyad, U. M., & Irani, K. B. (1992). On the handling of continuous-valued attributes in decision tree generation. *Machine Learning, 8*(1), 87–102.
10. Fayyad, U. M., & Irani, K. B. (1993). Multi-interval discretization of continuous-valued attributes for classification learning. In *Proceedings of the 13th International Joint Conference on Artificial Intelligence (IJCAI)* (pp. 1022–1029).

11. García, S., Luengo, J., & Herrera, F. (2015). *Data preprocessing in data mining*. New York: Springer.
12. García, S., Luengo, J., Sáez, J. A., López, V., & Herrera, F. (2013). A survey of discretization techniques: Taxonomy and empirical analysis in supervised learning. *IEEE Transactions on Knowledge and Data Engineering, 25*(4), 734–750.
13. Hu, H.-W., Chen, Y.-L., & Tang, K. (2009). A dynamic discretization approach for constructing decision trees with a continuous label. *IEEE Transactions on Knowledge and Data Engineering, 21*(11), 1505–1514.
14. Liu, H., Hussain, F., Tan, C. L., & Dash, M. (2002). Discretization: An enabling technique. *Data Mining and Knowledge Discovery, 6*(4), 393–423.
15. Machine Learning Library (MLlib) for Spark. (2019) MLlib. http://spark.apache.org/docs/latest/mllib-guide.html.
16. Parthasarathy, S., & Ramakrishnan, A. (2002). Parallel incremental 2D-discretization on dynamic datasets. In *International Conference on Parallel and Distributed Processing Systems* (pp. 247–254).
17. Pinto, C. (2006). Discretization from data streams: applications to histograms and data mining. In *In Proceedings of the 2006 ACM symposium on Applied computing (SAC06* (pp. 662–667).
18. Quinlan, J. R. (1993). *C4.5: programs for machine learning*. San Francisco, CA: Morgan Kaufmann Publishers Inc.
19. Ramírez-Gallego, S., García, S., Benítez, J. M., & Herrera, F. (2016). Multivariate discretization based on evolutionary cut points selection for classification. *IEEE Transactions on Cybernetics, 46*(3), 595–608.
20. Ramírez-Gallego, S., García, S., Benítez, J. M., & Herrera, F. (2018). A distributed evolutionary multivariate discretizer for big data processing on Apache spark. *Swarm and Evolutionary Computation, 38*, 240–250.
21. Ramírez-Gallego, S., García, S., & Herrera, F. (2018). Online entropy-based discretization for data streaming classification. *Future Generation Computer Systems, 86*, 59–70.
22. Ramírez-Gallego, S., García, S., Talín, H. M., Martínez-Rego, D., Bolón-Canedo, V., Alonso-Betanzos, A., et al. (2016). Data discretization: taxonomy and big data challenge. *Wiley Interdisciplinary Reviews: Data Mining and Knowledge Discovery, 6*(1), 5–21.
23. van Leeuwen, J., & Wood, D. (1993). Interval heaps. *The Computer Journal, 36*(3), 209–216.
24. Vitter, J. S. (1985). Random sampling with a reservoir. *ACM Transactions on Mathematical Software, 11*(1), 37–57.
25. Webb, G. I. (2014). Contrary to popular belief incremental discretization can be sound, computationally efficient and extremely useful for streaming data. In *Proceedings of the 2014 IEEE International Conference on Data Mining, ICDM '14* (pp. 1031–1036). Washington, DC: IEEE Computer Society.
26. Witten, I. H., Frank, E., Hall, M. A., & Pal, C. J. (2017). *Data mining: practical machine learning tools and techniques*. Cambridge, MA: Morgan Kaufmann Publisher.
27. Wu, X., & Kumar, V. (Eds.). (2009). *The top ten algorithms in data mining*. Chapman & Hall/CRC Data Mining and Knowledge Discovery. New York: CRC Press.
28. Wu, X., Zhu, X., Wu, G.-Q., & Ding, W. (2014). Data mining with big data. *IEEE Transactions on Knowledge and Data Engineering, 26*(1), 97–107.
29. Xu, Y., Wang, X., & Xiao, D. (2012). A two step parallel discretization algorithm based on dynamic clustering. In *Proceedings of the 2012 International Conference on Computer Science and Electronics Engineering - Volume 03, ICCSEE '12* (pp. 192–196).
30. Yang, Y., & Webb, G. I. (2009). Discretization for naive-Bayes learning: managing discretization bias and variance. *Machine Learning, 74*(1), 39–74.
31. Zhang, Y., Yu, J., & Wang, J. (2014) Parallel implementation of chi2 algorithm in MapReduce framework. In *International Conference on Human Centered Computing* (pp. 890–899). Heidelberg: Springer.

32. Zhao, Y., Niu, Z., Peng, X., & Dai. L. (2011). A discretization algorithm of numerical attributes for digital library evaluation based on data mining technology. In *Proceedings of the 13th International Conference on Asia-pacific Digital Libraries: For Cultural Heritage, Knowledge Dissemination, and Future Creation, ICADL'11* (pp. 70–76).
33. Zighed, D. A., Rabaséda, S., & Rakotomalala, R. (1998). FUSINTER: A method for discretization of continuous attributes. *International Journal of Uncertainty, Fuzziness and Knowledge-Based Systems, 06*(03), 307–326.

Chapter 8
Imbalanced Data Preprocessing for Big Data

8.1 Introduction

Among the wide set of problems worsened or even directly provoked by Big Data, we can highlight the treatment of imbalanced data in classification. Many today's problems possess classes represented by a negligible number of examples compared to the other classes. Moreover, the classes which are underrepresented are typically those that arouse most interest; therefore, its correct identification becomes primary. This scenario, known as imbalanced classification, has gained lots of attention in the last years [13, 19].

Classifying imbalanced datasets is not usually a trivial task. Standard learning techniques are often guided by global search measures that constantly overlook this contingency. Related studies asserted the power of classifiers suffer a dramatic drop in the event of both imbalance and lack of data [10]. In this manner, it is necessary to stress the main description features of problem and solve the drawbacks associated with them prior to the learning phase.

A large number of approaches have been conceived to address imbalanced classification. These approaches fall largely in two groups: data sampling solutions [2, 4], which modify the original training set, and algorithmic modifications [18] which alter standard algorithms in order to improve the prediction power on minority examples. Cost-sensitive solutions [7] combine the two previous options in order to minimize the error cost, usually higher in the minority side. In this chapter, we will focus on the former category where input data are preprocessed to equalize classes.

The techniques used to deal with Big Data are focused on obtaining fast, scalable, and parallel implementations. To achieve this goal, one of the more popular solutions nowadays is to follow a MapReduce procedure, dividing the original set into subsets which are easier to address, and then combining the partial solutions

© Springer Nature Switzerland AG 2020
J. Luengo et al., *Big Data Preprocessing*,
https://doi.org/10.1007/978-3-030-39105-8_8

obtained typically using an ensemble mechanism. However, data partition may have a special negative impact on imbalanced datasets. Among the drawbacks that degrade the performance in standard imbalanced classification, we can encounter the problem of small sample size and data lack [10] which are amplified by the data partitioning process performed in distributed systems. In these cases, oversampling tends to perform better than undersampling as this family of techniques mainly focus on increasing number of representative examples in maps.

Related to data lack, the small disjuncts [19] problem stands as one of the major problems in large-scale imbalanced learning. Small disjuncts are defined as those concepts that take form of small clusters of examples in the input space, also called "rare cases." The small disjuncts problem is also worsened by the division process in MapReduce jobs.

Other state-of-the-art oversampling methods such as SMOTE tend to fail when they are applied in distributed environments [8]. This poor performance may be caused by the random partitioning scheme introduced in each map. This division will provoke that some artificial samples are erroneously introduced on behalf of real objects with no spatial relationship. Novel SMOTE-based designs that introduce either a novel exact models (no voting), or a new partitioning scheme that allows relations between examples, are thus required.

In this chapter, we shall begin by analyzing a MapReduce framework for imbalanced data preprocessing with several state-of-the-art sampling techniques such as random undersampling and oversampling (Sect. 8.2). Section 8.3 presents a Big Data implementation of the popular SMOTE algorithm on Apache Spark. Section 8.4 describes a real imbalanced Big Data challenge and the winner algorithm. Section 8.5 enumerates the latest proposals on imbalanced Big Data preprocessing which rely on different techniques and strategies: ensemble voting, rough sets, or evolutionary learning, among others. Section 8.6 concludes this chapter with a summary of the section and some conclusions.

8.2   Data Sampling in Big Data

In many real-world supervised learning application, there is a difference between the amount of examples of the different classes. This difference leads to the learning algorithms to bias towards the most represented class. This situation is known as the class imbalance problem. One of the most extended techniques for dealing with imbalanced datasets is resampling.

In this section, we will describe the ROS-BD (Random OverSampling for Big Data) and RUS-BD (Random UnderSampling for Big Data) algorithms (Sects. 8.2.1 and 8.2.2, respectively) [6]. All of them are implemented under the MapReduce paradigm and designed to create a balanced version of the initial input dataset.

8.2.1 ROS-BD: Random Oversampling for Big Data

ROS algorithm balances the dataset by randomly replicating minority class instances from the original dataset until the number of instances from the minority and majority classes is the same. The ROS algorithm has been adapted to deal with Big Data environments by following a MapReduce design where each map process is responsible for adjusting the class distribution in each mapper through the random replication of minority class instances. On the other hand, the reduce process is responsible for collecting the outputs generated by each mapper to form the balanced dataset.

This process (along with RUS) is illustrated in Fig. 8.1. This consists of four steps: initial, map, reduce, and final. During the initial step, the algorithm performs a segmentation of the training set into independent data blocks and replicates and transfers them to other machines. Next, in the map step, each map task balances the class distribution through the random replication of minority class examples. Then, the reduce step collects the output generated by each mapper and randomizes the instances in the balanced dataset. In the final step, the balanced dataset that is generated in the reduce process forms the final dataset that will be the input data for the subsequent classifier.

Algorithms 1 and 2 give the pseudocode of the map and reduce functions of the MapReduce job, respectively. In Algorithm 1, Step (3) calculates the total number of replicas of each instance and is referred as the replication factor. For example, a replication factor of 1 means that there is only one copy of each instance in each mapper, a replication factor of 2 means two copies of each instance and so on. This replication factor is calculated with the total majority class instances and the total

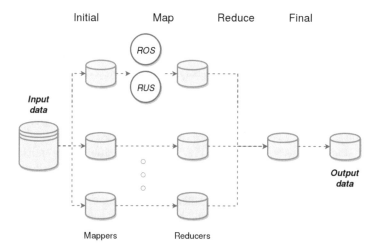

Fig. 8.1 A flowchart of the MapReduce design for all the sampling techniques (ROS, RUS). The output of each mapper will contain the local instances ordered by class label

---

Algorithm 1 Map phase for ROS-BD (key, value)

Input: (key, value) pair, where key is the offset in bytes and value is a given instance.
Output: (key', value') pair, where key' is the offset in bytes and value' is a given instance.
1: *instance ← instanceRepresentation(value)*
2: *class ← instance.getClass()*
3: *factor ← computeReplication(class)*
4: **for** *i = 0 to factor −* 1 **do**
5:   *EMIT(key, instance)*
6: **end for**

Algorithm 2 Reduce phase for ROS-BD (key, value)

Input: values, where values is a list of pairs.
Output: (key', value') pair, where key' is a null value and value' is a given instance.
1: **while** *values.hasNext()* **do**
2:   *instance ← instanceRepresentation(value.getValue())*
3:   *instances ← instances.add(instance)*
4: **end while**
5: *final ← randomize(instance)*
6: **for** *i = 0 to final.length −* 1 **do**
7:   *EMIT(null, final.get(i))*
8: **end for**

---

instances of the class of the instance that we want to replicate. Step (5) outputs the intermediate (key', value') pair (key, instance). When each mapper has finished, Algorithm 2 is called to randomize (Step 5) the final instances obtained previously and write them as final output (Step 7).

8.2.2 *RUS-BD: Random Undersampling for Big Data*

Contrary to ROS, RUS randomly deletes majority class examples from the original dataset until the balance with the minority class is achieved. In the RUS version adapted to Big Data, each map process is responsible for grouping by classes all the instances in its data partition and the reduce process is responsible for collecting the output by each mapper and equalizing the class distribution through the random elimination of majority class instances to form the balanced dataset.

This 4-steps process is illustrated in Fig. 8.1. Firstly, the algorithm splits the input dataset into independent data blocks; these blocks are then automatically replicated and transferred between the distinct cluster nodes. Then, each mapper processes and groups by classes all the local instances. Afterwards, the reduce step collects the output generated by each mapper and balances the class distribution randomly eliminating majority class examples. Finally, the balanced dataset that is generated in the reduce process is the final dataset that will be the input data for the subsequent classifier.

Algorithm 3 Map phase for RUS-BD (key, value)

Input: (key, value) pair, where key is the offset in bytes and value is a given instance.
Output: (key', value') pair, where key' is the class label associated to value and value' is a given
    instance.
1: *instance* ← *instanceRepresentation*(*value*)
2: *class* ← *instance.getClass*()
3: *EMIT*(*class, value*)

Algorithm 4 Reduce phase for RUS-BD (key, value)

Input: (key, value) pair, where key represents a class name and values is a list of instances.
Output: (key', value') pair, where key' is a null value and value' is a given instance.
1: **if** *key* == *majorityClass* **then**
2:     **while** *values.hasNext*() **do**
3:         *instance* ← *instanceRepresentation*(*value.getValue*())
4:         *instances* ← *instances.add*(*instance*)
5:     **end while**
6:     *instances* ← *shuffle*(*instances*)
7:     *instances* ← *instances.sublist*(0, #*minorityclass* − 1)
8:     *final.add*(*instances*)
9: **else**
10:     **while** *values.hasNext* **do**
11:         *instance* ← *instanceRepresentation*(*values.getValue*())
12:         *final.add*(*instance*)
13:     **end while**
14: **end if**
15: *final* ← *randomize*(*final*)
16: **for** *i* = 0 *to final.length* − 1 **do**
17:     *EMIT*(*null, final.get*(*i*))
18: **end for**

Algorithm 3 shows all the operations included in the map function, which generates a new (key, value) pair for each input pair. The new key is the class name of the input instance and the new value is the instance as such. Finally, the reduce function will receive all the instances grouped by class.

Algorithm 4 shows the pseudocode of the reduce function. In this algorithm, the Steps (2–8) are executed when the input key is equal to the majority class. In this case, the input values are shuffled (Step 6) to select (Step 7) a number of majority class instances equal to the number of minority class instances. In case of minority class, Steps (10–13), the algorithm maintains the input values as final instances. Once it is finished, a final process is called to randomize the instances obtained that later will fed the classifier.

In Fig. 8.2 we can find a Spark Package associated with ROS and RUS algorithms.

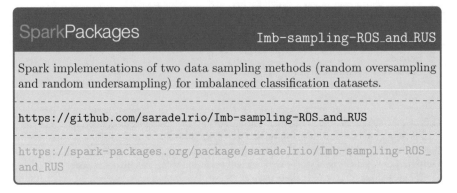

Fig. 8.2 Spark package: ROS-BD and RUS-BD

8.3 SMOTE-BD: SMOTE for Big Data

In this section, we will describe the SMOTE-BD algorithm (Synthetic Minority Oversampling TEchnique for Big Data), an exact and fully scalable SMOTE [4] in Spark for Big Data [1]. It is a Big Data implementation of the classic SMOTE algorithm. SMOTE forms new minority class examples by interpolating between several neighbor minority class examples.

First, the algorithm performs a filtering over the training set to get the minority and majority subsets of instances. Then, the minority data, which is partitioned according to an algorithm parameter, is normalized taking into account the statistics of the full training set and is cached to be reused in the following steps.

Later, nearest neighbors for each positive instance is obtained using an exact implementation of KNN in Spark (KNN_IS) [20] which splits the training dataset in a user-defined number of partitions, calculates for each instances in a chunk its neighbors and finally, in a reduction phase, makes a final list of k-nearest neighbors.

After that, the generation of artificial minority class instances is begun. All the nearest neighbors obtained in the previous step are broadcasted to the main memory of the all nodes in the cluster. The *broadcast* operation allows to keep a read-only variable cached on each node rather than shipping a copy to each task, and it performs this action in an efficient manner.

Then, for each positive instance in a data partition and using the broadcasted variable, the algorithm generates the corresponding number of synthetic examples by interpolating between each minority instance and its k-nearest neighbors. Figure 8.3 depict how to create synthetic data points in the SMOTE algorithm.

Finally, the algorithm performs a denormalization process over the artificial dataset and joins the original positive and negative instances with the artificial ones in order to conform the balanced dataset.

SparkPackages smote-bd

It is a fully scalable preprocessing approach for imbalanced classification in Big Data. It is based on one of the most widespread preprocessing solutions for imbalanced classification, namely the SMOTE algorithm, which creates new synthetic instances according to the neighborhood of each example of the minority class.

```
spark-shell --packages majobasgall:smote-bd:0.1
```

https://spark-packages.org/package/majobasgall/smote-bd

Fig. 8.3 Spark package: SMOTE-BD

Algorithm 5 SMOTE-BD algorithm

```
Input: Tr, Ts, ratio, k, nP, nR, nIt, minClassLabel
origData ← textFile(Tr)
minData ← origData.filter(labels == minClassLabel)
minData ← minData.map(normalize).repartition(nP)
neighbors ← KNN_IS.setup(Tr, Tr, k, nR, nI).calculatekNeighbours()
crFactor ← (nMajnMin)/nMin
neighbors ← broadcast(neighbors)
balancedData ← null
synData ← null
for i < nIt do
    synTmp ← minData.mapPartitionsWithIndex(
            createSynthData(index, partData, neighbors, crFactor, k))
    if synData == null then
        synData ← synTmp
    elsesynData ← synData.union(synTmp)
    end if
end for
synData ← synData.map(denormalize)
balancedData ← synData.union(origData)
return (balancedData)
```

Algorithms 5 and 6 show a pseudocode of the sequence of actions described above. The former covers the main program and the latter the function to create each artificial instance. This function invokes another function called interpolation which is in charge of doing the interpolation between two points. There is no pseudocode of this due to its simplicity.

In Fig. 8.3 we can find a Spark Package associated with the SMOTE algorithm.

Algorithm 6 Function to create synthetic instances between the minority class examples and their neighbors

---

1: $articialData \leftarrow null$
2: **for** $i < nIt$ **do**
3:      $firstInstance \leftarrow partitionData; nc = 0tocrFactor$
4:      $selNeighbor \leftarrow newRandom().nextInt(k)$
5:      $secondInstance \leftarrow neighbors(selNeighbor)$
6:      $newIntance \leftarrow interpolation(firstInstance, secondInstance)$
7:      $artificialData.add(newInstance)$
8: **end for**
9: **return** $(articialData)$

8.4   Imbalanced Big Data Competition

In this section, we will describe the ROSEFW-RF algorithm (Random OverSampling and Evolutionary Feature Weighting for Random Forest), the winner algorithm for the ECBDL'14 big data competition [24]. This algorithm utilized random oversampling in a real Big Data contest, called ECBDL'14 Big Data challenge, to score first. The problem analyzed in this contest consists of 631 features and 32 millions of instances with an imbalance rate of 2%. The winner algorithm was formed by several MapReduce stages combined to provide a final solution. These stages have different purposes that range from: (1) equalizing class distribution, (2) selecting the most relevant features using a feature weighting process [21], (3) learning from a Random Forest model built on preprocessed data, (4) and finally predicting unlabeled data.

The two main outcomes derived from this study are: the remarkable effectiveness of oversampling (and specifically, ROS-BD) when applied to extremely imbalanced large-scale problems like that addressed in ECBDL'14 [5]; and the goodness of the preprocessing scheme presented where feature selection and imbalanced preprocessing join forces to tackle the huge imbalance ratio. The main drawback showed in [24] is related to the high oversampling ratio needed to equalize distributions, which makes the final solution highly time-consuming. In this method, each map task performed a whole evolutionary feature weighting cycle in its data partition and emitted a vector of weights. Then, the reduce process was responsible of the iterative aggregation of all the weights provided by the maps. Finally, the resulting weights were used with a threshold to select the most important characteristics.

In the ECBDL'14 big data challenge three metrics were used to assess the prediction results: true positive rate (TPR: TP/P), true negative rate (TNR: TN/N), accuracy, and the final score of TPR · TNR. The final score was chosen because of the huge class imbalance of the dataset in order to reward methods that try to predict well the minority class of the problem.

Table 8.1 collects the results obtained with the five different approximations. The initial experiment uses a 100% of oversampling ratio and RF as classifier. The second experiment consisted on increasing the oversampling ratio to further

Table 8.1 Results obtained with the different approximations

Method	#Maps	TPR	TNR	TPR · TNR
ROS 100% + RF	192	0.580217	0.821987	0.476931
ROS 130% + RF	64	0.670189	0.758622	0.508420
ROS 130% + RF + FW 90	64	0.674754	0.777440	0.524580
ROS 130% + RF 25 + FW 90	64	0.669531	0.784856	0.525486
ROS 170% + RF 25 + FW 90	64	0.730432	0.730183	0.533349

balance TPR and TNR. The third approximation was to detect relevant features via evolutionary featuring weighting (90 features selected). The fourth approach was focused on the RF parameters. Instead of using the $\log(\#Features)+1$, that resulted in 8 features, authors incremented to 25. The final, and winner, solution consisted on increasing the oversampling rate to perfectly balance the TPR and TNR.

In this particular problem, the necessity of balancing the TPR and TNR ratios emerged as a difficult challenge for most of the participants of the competition. In this sense, the results of the competition have shown the goodness of the proposed MapReduce methodology. Particularly, the modular ROSEFW-RF methodology composed of several, highly scalable, preprocessing and mining methods has shown to be very successful in this challenge and outperform the other participants.

8.5   Other Studies on Imbalance Data Preprocessing for Big Data

As happened in other preprocessing fields, just few proposals have been presented in the literature able to process large-scale datasets. Most of them are distributed implementation of golden standard algorithms, such as random oversampling/undersampling or SMOTE-based methods. Although the ensemble model is the prevailing trend in these techniques, there also exist some ad hoc models that range from evolutionary undersampling to rough sets based SMOTE.

8.5.1   Evolutionary Undersampling

In order to solve the problems presented by large-scale undersampling in [6], Triguero et al. presented a work [23] that offer a parallel solution based on evolutionary learning. This undersampling MapReduce algorithm is formed by two complete stages. This former one undersamples data and then learns a decision tree on resulting data, whereas the latter one aims at labeling test data.

As a second level parallelization, a windowing scheme can be introduced in the mappers. Windowing serves here to reduce the runtime associated with the fitness evaluation. The subset in each map thread is then divided into several disjoint stratas. In each iteration, only one stratum is used to evaluate the population, which changes in the subsequent iterations following a round-robin policy.

In order to test the performance of the previous method, authors designed an experimental framework that includes C4.5 applied on different versions of the KDDCup'99 datasets. Results confirmed the appropriateness of the global model both in terms of precision and runtime. An extension to this model implemented under the Spark framework was recently presented in [22].

8.5.2   KNN Based Data Cleaning

In [16], authors propose an alike scheme to equilibrate class proportions in DNA large-scale problems. Concretely, Kamar et al. created a distributed clustering methodology using data reduction with K-nearest neighbors (KNN).

This model, specially designed for large-scale bioinformatics problems, was able to process up to 90 million pairs in the experiments. Furthermore, authors compare the performance of KNN on the reduced version versus the original dataset, as well as provide a complete scalability analysis.

8.5.3   NRSBoundary-SMOTE

A MapReduce implementation of Neighborhood RoughSet Theory [14] (NRSBoundary-SMOTE) is presented in [15]. This ad hoc version is based on two MapReduce procedures: one that partitions the original data, and another one that oversamples the minority class. At a more detailed level, the first phase split the training set into three different subsets (positive, minority, and boundary) following a neighborhood relation. *Positive* consists of the majority class examples whose neighborhood has the sample class label. The *minority* set is the simplest one, and contains the minority examples. Finally, *boundary* filters those minority examples with any majority class examples present in its neighborhood. In the second stage, mappers fetch blocks from the boundary set and locally compute the neighborhood for each instance. Afterwards, reducers randomly select for each sample one of its neighbors. These selected examples will be used for interpolation purposes. If the new generated example belongs to the any neighborhood in positive, another neighbor will be selected from list. Else the synthetic example will be maintained.

The positive and minority sets are both allocated in the Hadoop Distributed Cache in order to speed up read/write accesses. However, this feature undermines the scalability of NRSBoundary-SMOTE as it requires the entire set to be cached.

8.5.4   Ensemble ELM with Resampling

Based on MapReduce and ensemble learning mechanism, Zhai et al. propose a binary classification algorithm for imbalanced Big Data datasets [25]. It consists of four stages: (1) alternately oversample between positive and negative instances; (2) construct several balanced data subsets based on the novel positive class instances; (3) train several component classifiers using the extreme learning machine (ELM) algorithm on the previous subsets; (4) integrate the ELM classifiers via simple voting.

Concretely, the first stage starts by computing the center of positive instances. Then, it samples instances along the line that connects the center and each positive instance. Next, for each instance generated, its k-nearest neighbors with negative class are retrieved using MapReduce. Afterwards, the algorithm samples instances along the line between the given instance and its k-nearest negative neighbors. This oversampling process is repeated p times.

In the second stage, the same amount of instances are sampled from the negative class as generated for the positive class. This process is repeated l times. Each round, the resulting instance from both classes is coalesced to obtain a balanced subset.

Authors expose that their solution has two advantages with respect to the state-of-the-art: (1) it can extend the learning region belonging to the positive class instances, (2) it is able to classify imbalanced large datasets thanks to the MapReduce scheme followed. Furthermore, the experimental outcomes presented in this work show that the proposal outperforms the other alternatives (SMOTE-Vote, SMOTE-Boost, and SMOTE-Bagging) in terms of test accuracy and time performance.

8.5.5   Imbalance Treatment for Multiclass Problems

In [3], authors offer a preliminary study on multiclass imbalanced classification for Big Data. The solution proposed is a scalable one-vs-all binarization algorithm for Spark and Hadoop platforms.

This algorithm is formed by two stages. Firstly, a One-vs-All process is used to split the original datasets into several subsets with two classes (sequential process). Then, an instance of SMOTE-BD [6] is launched on each binary subset in order to equalize class distribution, following the scheme proposed in [9]. Lastly, Mahout's Random Forest is elected to train the final model and perform predictions.

This work is interesting as a first approach on the field, but it lacks from a real distributed binarization technique and a large enough experimental framework where datasets exceed millions of examples.

8.5.6   SMOTE for GPU

A GPU-based extension to SMOTE-BD was proposed in [12], which relies on a smart utilization of main memory to adapt the previous idea to parallel computation. The two main improvements introduced in this version are: a novel GPU implementation of supporting KNN algorithm [11], and the only inclusion of minority class in memory.

A further extension for high-speed data streams was thought in [17]. Here an extreme machine learning algorithm for GPU among with undersampling and oversampling was used to predict labels in dynamic environments. Krawczyk et al. show that the utilization of sampling techniques reduces the negative impact of skewed distributions on the learner's performance and it adapts better to non-stationary streams.

8.6   Summary and Conclusions

Imbalance learning is one of today most common scenarios. This makes the classifier focus on the most represented classes, forgetting those that are less represented in the data.

In this chapter we have provided several solutions for these problem. The most popular and widely used algorithms are ROS and RUS. These methods balance the data by repeating the minority class (ROS) or by removing instances from the majority class (RUS). Most advanced methods like SMOTE are also presented, which generates artificial instances for balancing the data. We have also presented a real-world imbalanced Big Data competition, and the winner algorithm based on data balancing.

Although we have analyzed several proposals for imbalanced data preprocessing, there is still little research devoted to tackle this problem in Big Data environments, in comparison with normal-sized data. Many open challenges arise from the study of the imbalanced learning state in Big Data scenarios. The main challenge is the thorough design at the implementation level for current algorithms to address Big Data problems. Also, there is a need for novel algorithms for the generation of artificial instances. One final open challenge will be to analyze the ratio between classes.

References

1. Basgall, M. J., Hasperué, W., Naiouf, M., Fernández, A., & Herrera, F. (2018). SMOTE-BD: An exact and scalable oversampling method for imbalanced classification in big data. *Journal of Computer Science and Technology, 18*(03), e23.
2. Batista, G. E. A. P. A., Prati, R. C., & Monard, M. C. (2004). A study of the behavior of several methods for balancing machine learning training data. *ACM SIGKDD Explorations Newsletter, 6*(1), 20–29.

3. Bhagat, R. C., & Patil, S. S. (2015). Enhanced smote algorithm for classification of imbalanced big-data using Random Forest. In *Souvenir of the 2015 IEEE International Advance Computing Conference, IACC 2015* (pp. 403–408)

4. Chawla, N. V., Bowyer, K. W., Hall, L. O., & Kegelmeyer, W. P. (2002). SMOTE: Synthetic minority over-sampling technique. *Journal of Artificial Intelligence Research, 16*, 321–357.

5. del Río, S., Benítez, J. M., & Herrera, F. (2015). Analysis of data preprocessing increasing the oversampling ratio for extremely imbalanced Big Data classification. In *2015 IEEE Trustcom/BigDataSE/ISPA* (Vol. 2, pp. 180–185).

6. del Río, S., López, V., Benítez, J. M., & Herrera, F. (2014). On the use of MapReduce for imbalanced Big Data using random forest. *Information Sciences, 285*, 112–137.

7. Elkan, C. (2001). The foundations of cost-sensitive learning. In *In Proceedings of the Seventeenth International Joint Conference on Artificial Intelligence* (pp. 973–978).

8. Fernández, A., del Río, S., Chawla, N. V., & Herrera, F. (2017). An insight into imbalanced big data classification: Outcomes and challenges. *Complex & Intelligent Systems, 3*(2), 105–120.

9. Fernández, A., López, V., Galar, M., Del Jesus, M. J., & Herrera, F. (2013). Analysing the classification of imbalanced data-sets with multiple classes: Binarization techniques and ad-hoc approaches. *Knowledge-Based Systems, 42*, 97–110.

10. Guo, Y., Graber, A., McBurney, R. N., & Balasubramanian, R. (2010). Sample size and statistical power considerations in high-dimensionality data settings: A comparative study of classification algorithms. *BMC Bioinformatics, 11*, 447.

11. Gutierrez, P. D., Lastra, M., Bacardit, J., Benitez, J. M., & Herrera, F. (2016). GPU-SME-kNN: Scalable and memory efficient kNN and lazy learning using GPUs. *Information Sciences, 373*, 165–182.

12. Gutierrez, P. D., Lastra, M., Benitez, J. M., & Herrera, F. (2017). SMOTE-GPU: Big data preprocessing on commodity hardware for imbalanced classification. *Progress in Artificial Intelligence, 6*(4), 347–354.

13. He, H., & Garcia, E. A. (2009). Learning from imbalanced data. *IEEE Transactions on Knowledge and Data Engineering, 21*(9), 1263–1284.

14. Hu, F., & Li, H. (2013). A novel boundary oversampling algorithm based on neighborhood rough set model: NRSBoundary-SMOTE. *Mathematical Problems in Engineering, 2013*, 1–10.

15. Hu, F., Li, H., Lou, H., & Dai, J. (2014). A parallel oversampling algorithm based on NRSBoundary-SMOTE. *Journal of Information and Computational Science, 11*(13), 4655–4665.

16. Kamal, S., Ripon, S. H., Dey, N., Ashour, A. S., & Santhi, V. (2016). A MapReduce approach to diminish imbalance parameters for big deoxyribonucleic acid dataset. *Computer Methods and Programs in Biomedicine, 131*, 191–206.

17. Krawczyk, B. (2016). GPU-accelerated extreme learning machines for imbalanced data streams with concept drift. In M. Connolly (Ed.), *The International Conference on Computational Science, Procedia Computer Science* (Vol. 80, pp. 1692–1701)

18. López, V., Fernández, A., del Jesus, M. J., & Herrera, F. (2013). A hierarchical genetic fuzzy system based on genetic programming for addressing classification with highly imbalanced and borderline data-sets. *Knowledge-Based Systems, 38*, 85–104. Special Issue on Advances in Fuzzy Knowledge Systems: Theory and Application.

19. López, V., Fernández, A., García, S., Palade, V., & Herrera, F. (2013). An insight into classification with imbalanced data: Empirical results and current trends on using data intrinsic characteristics. *Information Sciences, 250*, 113–141.

20. Maíllo, J., Ramírez, S., Triguero, I., & Herrera, F. (2017). kNN-IS: An Iterative Spark-based design of the k-nearest neighbors classifier for big data. *Knowledge-Based Systems, 117*, 3–15.

21. Triguero, I., Derrac, J., García, S., & Herrera, F. (2012). Integrating a differential evolution feature weighting scheme into prototype generation. *Neurocomputing, 97*, 332–343.

22. Triguero, I., Galar, M., Merino, D., Maillo, J., Bustince, H., & Herrera, F. (2016). Evolutionary undersampling for extremely imbalanced big data classification under apache spark. In *IEEE Congress on Evolutionary Computation (CEC 2016)*, Vancouver (pp. 640–647).

23. Triguero, I., Galar, M., Vluymans, S., Cornelis, C., Bustince, H., Herrera, F., & Saeys, Y. (2015). Evolutionary undersampling for imbalanced Big Data classification. In *2015 IEEE Congress on Evolutionary Computation (CEC)* (pp. 715–722).
24. Triguero, I., Río, S., López, V., Bacardit, J., Benítez, J. M., & Herrera, F. (2015). ROSEFW-RF: The winner algorithm for the ECBDL'14 Big Data competition: An extremely imbalanced Big Data bioinformatics problem. *Knowledge-Based Systems, 87*, 69–79.
25. Zhai, J., Zhang, S., & Wang, C. (2015). The classification of imbalanced large data sets based on MapReduce and ensemble of elm classifiers. *International Journal of Machine Learning and Cybernetics*, 1–9.

Chapter 9
Big Data Software

9.1 Introduction

Big Data Analytics is nowadays sitting at the forefront of many disciplines that are not directly related to computer science, statistics, or maths. The advent of the Internet of Things, the Web 2.0, and the great advances in technology are transforming many areas such as medicine, business, transportation, or energy by collecting massive amounts of information [1, 12, 16, 30]. The impact of exploiting this data may reflect on competitive advantages for companies or unprecedented discoveries in multiple science fields [27]. Nevertheless, both companies and researchers are facing major challenges to cope with the Volume, Velocity, Veracity, and Variety (among others V's) that characterize this flood of data. These V's define the main issues of the Big Data problem [15].

With the leverage of distributed technologies such as the MapReduce programming paradigm and the Apache Spark platform [13, 50], some classical data mining (DM) algorithms are being adapted to this new data-intensive scenario [22, 29]. However, Big Data mining techniques are not only confronted with scalability or speed issues (volume/velocity) and they will also have to handle inaccurate data (noisy or incomplete) and massive amounts of redundancy [43].

As stated before, Big Data is typically characterized by a Volume, Velocity, Variety, and Veracity (among other V's) that poses a challenge for current technologies and algorithms. The problem of Big Data has many different faces such as data privacy/security, storage infrastructure, visualization, or analytics/mining. Tackling big datasets with DM and machine learning (ML) algorithms means moving from sequential to distributed systems that can make use of a network of computers to operate faster. However, parallel computation has been around for many years, what is then new with Big Data? *"The principle of data locality."* Traditional HPC systems have provided a way to accelerate computation by means of parallel programming models such as MPI [39]. Classical HPC systems fail to scale out when data-intensive applications are involved, as data will be moved

© Springer Nature Switzerland AG 2020
J. Luengo et al., *Big Data Preprocessing*,
https://doi.org/10.1007/978-3-030-39105-8_9

across the network causing significant delays. In a Big Data scenario, minimizing the movement of data across the network by keeping data locally in each computer node is key to provide an efficient response. Thus, ideally, each computer node will operate only on data that is locally available.

The MapReduce functional programming paradigm [13] and its open-source implementation in Hadoop [47] were the precursor parallel processing tools to tackle data intensive applications, by implementing the data locality principle. Hadoop implements this MapReduce programming model together with a distributed file systems that provides data locality across a network of computing nodes. As a result, the end-user is able to design scalable/parallel algorithms in a transparent way, so that data partitioning, job communication, and fault-tolerance are automatically handled. Despite the great success of Hadoop, researchers in the field of DM found serious limitations when consecutive operations needed to be applied on the same (big) data, reporting a significant slow-down. Many other frameworks have been made available to address these limitations of Hadoop, and one of the most popular platform nowadays is Apache Spark [50]. As a data processing engine, Spark operates with MapReduce-like functions on a distributed dataset, known as Resilient Distributed Dataset (RDD), which can be cached in main memory to allow for multiple iterations. Spark is evolving very quickly and more efficient APIs such as DataFrames and Datasets are being developed. On the other hand, Apache Flink [4] appeared as a data processing engine focused on Big Data streaming applications.

In order to extract all valuable knowledge from the data, several techniques need to be applied, like statistical tests, correlations, or normalization, among others. Big Data Analytics requires an even deeper analysis of the data because of its size. Classical ML methods cannot tackle this amount of data, as they were not conceived for the scalability problem. For these reasons, Big Data frameworks include a thorough list of algorithms ready to be applied to the data in form of ML libraries. This enables the practitioners to quickly obtain a first idea of the data distribution.

Here we describe and classify the two most popular ML libraries for both batch, and stream data processing. Firstly, we introduce Apache Spark MLlib and depict all of its included algorithms (Sect. 9.2). Next, we describe in depth BigDaPSpark, a Big Data library for data preprocessing under Apache Spark (Sect. 9.3). Then, FlinkML is described and studied in depth (Sect. 9.4). BigDaPFlink is introduced in Sect. 9.5, a Big Data streaming library focused on data preprocessing for Apache Flink. Finally, Sect. 9.6 summarizes the chapter and gives some conclusions.

9.2 MLlib: A Spark Machine Learning Library

Spark MLlib [28] is a powerful library of ML algorithms and utilities designed to run in parallel on a Spark cluster. MLlib enables the use of Spark in the data analytics field. The development of MLlib started in the UC Berkeley AMPLab in 2012, and was open-sourced in 2013. The first version of MLlib was package with Spark 0.8 release.

MLlib mission is to make practical ML easy and scalable. It should be easy to build ML applications and it should be capable of learning from large-scale real-world datasets. MLlib is not just a collection of algorithms, it contains the most popular methods and utilities in the ML ecosystem, along with an extensive documentation, making it a very complete toolbox for developers.

MLlib is capable of working with hundreds of data sources, including those belonging to the Hadoop Ecosystem, like HDFS. It is also able to work with data from relational databases, local data, or images. MLlib introduces a few new data types:

- Local vector: composed of 0-based indices and double-typed values. They are stored on a single machine. There are two types of local vectors: *dense* and *sparse*. A *dense* vector is formed by an array of double values, while the *sparse* vector is composed of two parallel arrays with its indices and values.
- Labeled point: it is the main data type of MLlib. It is composed of a local vector (*dense* or *sparse*) with a label. The label must be a double value, starting in 0 in the case of classification problems.
- Local matrix: like local vectors, local matrices are composed of double-typed values stored on a single machine. They can be both *dense* or *sparse*.
- Distributed matrix: they are stored distributively in one or more RDD. There are four types of distributed matrices implemented so far.

  The main type is the *RowMatrix*. It is a distributed matrix backed by an RDD of its rows, being each row a local vector. The *IndexedRowMatrix* is similar to a *RowMatrix* but with row indices. These indices are useful for identifying rows and joins operations. A *BlockMatrix* is another distributed matrix backed by an RDD of *MatrixBlock* which is a tuple of (Integer, Integer, Matrix). Finally, the *CoordinateMatrix* is stored in coordinate list (COO) format, and its backed by an RDD of its entries.

MLlib is currently formed by two packages, *mllib* and the most recent *ml*:

- *mllib*: this is the first version of MLlib, which was built on top of RDD. It contained the majority of the methods proposed in Spark up to 2.0 version.
- *ml*: it comes with the newest features of MLlib for constructing ML pipelines. This higher-level API is built on enhanced DataFrames structures [5]. It has become the primary ML API for Spark as of version 2.0.

Since Spark's *mllib* library does not include new algorithms as of version 2.0, we will focus on the *ml* library. Here, we describe and classify all DM techniques for Spark's *ml* ML library.[1]

---

[1]*MLlib* and *ml* documentation: http://spark.apache.org/docs/latest/ml-guide.html.

9.2.1   *Pipelines*

MLlib moves beyond a list of algorithms and help users develop their own ML workflows. Pipelines allow to chain multiple algorithms into a single workflow. They consist on a sequence of stages that need to be run in a specific order. This simplifies the execution of several transformations on a DataFrame. Pipelines were inspired by the scikit-learn project [10].

Pipelines are composed of two different main components:

- Transformers: converts a DataFrame into another DataFrame, typically adding one or more columns. An example of Transformer is a learning model that predicts each vector from the DataFrame, and appends the predicted label as a new column.
- Estimators: abstract the concept of learning algorithm or any algorithm that trains on data. They learn from a DataFrame and produce a model, which is a Transformer.

9.2.2   *Feature Extractors*

Feature extraction techniques aim for obtaining a set of features from raw data. These techniques are mainly used in text mining.

- TF_IDF: provides a numeric measure of how relevant is a word for a document. It is frequently used in text mining. The TF_IDF value increases proportionally to the number of times a word appears in the document, but is compensated by the frequency of the word in the document collection, which allows for handling the fact that some words are generally more common than others.
- Word2Vec: computes distributed vector representation of words. Similar words will be close in the vector space.
- CountVectorizer: converts a collection of text documents to vector of token counts.
- FeatureHasher: it projects features into indices in a vector. This is done by mapping features to indices in the feature vector.

9.2.3   *Feature Transformers*

Feature transformation techniques combine the original set of features to obtain a new set of less-redundant variables [23]. For example, by using projections to low-dimensional spaces.

- Binarizer: discretizes a set of features to binary (0/1) features using a given threshold.
- Single Value Decomposition (SVD): is a matrix factorization method that transform a real/complex matrix M ($m \times n$) into a factorized matrix A. The creators expose that for large matrices it is not needed the complete factorization but only to maintain the top-k singular values and vectors. In such way, the dimensions of the implied matrices will be reduced. They also assume that n is much smaller than m (tall-and-skinny matrices) in order to avoid a severe degradation of the algorithm's performance.
- PCA: tries to find a rotation such that the set of possibly correlated features transforms into a set of linearly uncorrelated features. The columns used in this orthogonal transformation are called principal components. This method is also designed for matrices with a low number of features.
- Polynomial Expansion: expands the set of features into a polynomial space. This new space is formed by an n-degree combination of the original dimensions.
- StringIndexer: encodes the labels of a string column into a new column of indices.
- IndexToString: transforms a column of label indices back to the original column, containing the labels as strings.
- VectorIndexer: indexes categorical features in datasets according to the number of categories of a feature.
- Interaction: takes two columns of vectors and generates a new column containing the product of all combinations of the values from each column.
- Normalizer: transforms a dataset by normalizing each example to have unit norm.
- StandardScaler: centers the data to have zero mean and/or scales the data to unit standard deviation.
- MinMaxScaler: scales each column to a given range (usually [0, 1]).
- MaxAbsScaler: scales the data to $[-1, 1]$ range by dividing through the maximum absolute value in each feature.
- Bucketizer: converts a column of continuous values into a column of feature bins. The thresholds are given by the user.
- ElementwiseProduct: scales each vector by a specified weight values using element-wise multiplication.
- VectorAssembler: combines a set of features into a single vector column.
- QuantileDiscretizer: discretizes each continuous feature into a categorical column. The number of thresholds is specified by the user.
- Imputer: fills the missing values (MV) in the dataset. It can use either the mean of the columns or the median.
- Locality-Sensitive Hashing (LSH): is aimed at hashing data points into bins, so as the data points close in the input domain will be placed in the same bins with high likelihood. LSH can be used for feature transformation by putting hashed values in new columns.

9.2.4 Feature Selectors

As explained before, feature selection (FS) tries to select relevant subsets of relevant features without incurring much loss of information [9].

- VectorSlicer: the user selects manually a subset of features.
- RFormula: selects features specified by an R model formula.
- Chi-Squared selector: it orders categorical features using a Chi-Squared test of independence from the class. Then, it selects the most-dependent features. This method can be categorized as univariate and filter, which needs to specify the number of selected features as input parameter.

9.3 BigDaPSpark

BigDaPSpark is a Big Data library focused on static data preprocessing [19], built on top of Apache Spark. This library is born with the objective of improving the Big Data ecosystem with new algorithms for Big Data preprocessing, in order to achieve Smart Data.

This library is composed of a series of algorithms for Big Data preprocessing under the Apache Spark framework. It contains algorithms for FS, data reduction, noise filtering, MV imputation, discretization, and imbalanced learning, among others. In Fig. 9.1 we can find the algorithm implementations associated with this library.

9.3.1 Feature Selection

The library contains a FS framework, implemented in a distributed fashion. It contains multiple information-theory based FS algorithms, like mRMR, InfoGain,

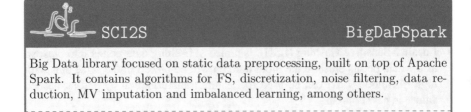

Fig. 9.1 BigDaPSpark library

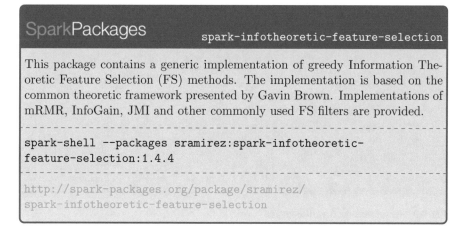

Fig. 9.2 Spark package: information theoretical FS framework

JMI and other commonly used FS filters [34]. This framework is described in Sect. 4.4. In Fig. 9.2 we can find a Spark Package associated with this research.

9.3.2 Data Reduction

The library contains four algorithms for performing data reduction based on the KNN algorithm: FCNN_MR, SSMASFLSDE_MR, RMHC_MR, and MR_DIS. As stated previously, the purpose of these algorithms is to obtain a reduced set of the original data that represents it as perfectly as possible. Some of these algorithms are implemented using a distributed framework, named MRPR [44]. This framework enables the use of iterative algorithms in Big Data environments by partitioning the input data in several chunks, and applying the corresponding algorithm independently to each one of them. After that process is finished, all the partitions are joined together using different strategies. These algorithms are described in depth in Sect. 5.4. In Fig. 9.3 we can find all the implementations of the techniques described in this section, available as a Spark Package.

- FCNN_MR: this algorithm is one of the most extended and widely used in data reduction [3]. It is an order-independent algorithm, based on the NN rule, to find a consistent subset of the training dataset. It has a quadratic time complexity in the worst-case. It also have showed to scale well on large and multidimensional datasets.
- SSMASFLSDE_MR: this algorithm is a hybrid and evolutionary algorithm composed of two methods. The first one is a steady-state memetic algorithm (SSMA) [18] that selects the most representative instances of the training set,

```
SparkPackages                                           SmartReduction
```

This framework implements four distance based Big Data preprocessing
algorithms for prototype selection and generation: FCNN_MR, SSMAS-
FLSDE_MR, RMHC_MR, MR_DIS, with special emphasis in their scalability
and performance traits.
- -
```
spark-shell --packages djgarcia:SmartReduction:1.0
```
- -
https://spark-packages.org/package/djgarcia/SmartReduction

Fig. 9.3 Spark package: SmartReduction

while the second one improves this subset by modifying the values of the selected
instances with a scale factor local search in differential evolution (SFLSDE)[42].
- RMHC_MR: random mutation hill climbing (RMHC) is a powerful yet simple
  algorithm for data reduction [38]. It starts by selecting a random sample of the
  data S. Then it randomly replaces an instance of the sample with one of the
  original data S^*. Next it uses both samples to calculate the classification accuracy
  in the complete dataset, using the KNN algorithm. The sample with the best
  accuracy is kept for the next iteration, were another instance will be substituted.
  After a determined number of iteration, the best sample is chosen.
- MR_DIS: is a parallel implementation of the democratic instance selection (IS)
  algorithm [6]. This algorithm applies a classic IS algorithm over an equally
  partitioned training dataset. The selected instances receive a vote. After a
  determined number of rounds, instances with most votes are removed from
  the data.

9.3.3   Noise Filtering

This section of the library is composed of two sub-libraries. The first one contains
three algorithms for removing noise in Big Data datasets: Homogeneous Ensemble
(HME-BD), Heterogeneous Ensemble (HTE-BD), and ENN-BD. These algorithms
are based on ensembles of classifiers, they were originally proposed in [20]. In
Fig. 9.4 we can find a Spark package associated with this research.

- HME-BD is based on a partitioning scheme of the dataset. It performs a k-fold of
  the input data, splitting the data into k partitions. The test partition is an unique
  1kth of the fold, and the train is the rest of the partition. Then it learns a deep
  Random Forest (a Random Forest with deep trees) in each fold, using the train

┌───┐
│ SparkPackages   NoiseFramework   │
├───┤
│ In this framework, two Big Data preprocessing approaches to remove   │
│ noisy examples are proposed: an homogeneous ensemble (HME_BD) and   │
│ an heterogeneous ensemble (HTE_BD) filter. A simple filtering   │
│ approach based on similarities between instances (ENN_BD) is also   │
│ implemented.   │
│ ─   │
│ spark-shell --packages djgarcia:NoiseFramework:1.2   │
│ ─   │
│ https://spark-packages.org/package/djgarcia/NoiseFramework   │
└───┘

Fig. 9.4 Spark package: NoiseFramework

partition as input. Once the learning process is finished, each of the k models learned predict the corresponding test partition of each fold. That way, the models will predict the data that they did not see while they were learned. The final step is to remove the noisy instances. This is done by a comparison of the original test labels with the predicted by the learners. If the labels are different, the instance is considered as noisy and removed. Finally, all the filtered partitions are joined together to compose a dataset clean of noise.

- HTE-BD shares the same workflow as HME-BD, but instead of using a unique classifier, it uses three of them. HTE-BD partitions the data performing a k-fold of the input data the same way as was described in HME-BD. Then it learns a deep Random Forest, a logistic regression, and a 1NN. With the predictions of the three models, a voting strategy is used to determine if an instance is noisy. There are two strategies available, *majority* and *consensus*. With the former, only two classifiers have to agree to take a decision. With the second, all classifiers must agree to consider an instance as noisy. The filtered partitions are joined to recompose the dataset without noise.

- ENN-BD is much simpler that the previous two. It is based on the similarity between instances [48]. It performs a KNN (typically $k = 1$ or $k = 3$) to the input data, and uses that same input data for prediction. That way, the closest neighbors for each instance are found. In order to remove the noisy instances, those neighbors are compared with the instance. If the label of the neighbors differs from the original, the instance is removed.

The second part of the noise library consists of three algorithms for noise filtering based on KNN [43]: AllKNN_BD, NCNEdit_BD and RNG_BD. In Fig. 9.5 we can find this software available as a Spark Package.

- AllKNN_BD: this method shares the same working scheme as ENN-BD with some exceptions. Instead of learning a 1NN, it learns several times KNN with different values of k (typically 1, 3 and 5) [41]. Each iteration it removes the

```
SparkPackages                                              SmartFiltering

This framework implements four distance based Big Data preprocessing algo-
rithms to remove noisy examples: ENN_BD, AllKNN_BD, NCNEdit_BD and
RNG_BD filters, with special emphasis in their scalability and performance
traits.
------------------------------------------------------------------------
spark-shell --packages djgarcia:SmartFiltering:1.0
------------------------------------------------------------------------
https://spark-packages.org/package/djgarcia/SmartFiltering
```

Fig. 9.5 Spark package: SmartFiltering

instances that does not agree with its closest neighbors. As can be expected, it is
a much aggressive noise filter than ENN-BD, as it applies KNN repeatedly.

• NCNEdit_BD: this algorithm uses the *k* nearest centroid neighborhood classification rule with the leave-one-out error estimate [35]. It discard instances if it is
  misclassified using the kNCN classification rule. In the NCN classification rule,
  the neighborhood is not only defined by the proximity of prototypes to a given
  instance, but also for their *symmetrical distribution* around it.

• RNG_BD: this noise filter computes the proximity graph of the data [36]. Then,
  all the graph neighbors of each instance give a vote for its class. If the label differs
  from the original label, the instance is considered as noise and removed.

These noise filtering algorithms are depicted in depth in Sects. 6.3 and 6.4,
respectively.

9.3.4   Missing Values Imputation

The library also contains two approaches, a global and a local implementation,
for MVs imputation using the k-nearest neighbor imputation, k-nearest neighbor—
local imputation, and k-nearest neighbor imputation—global imputation [43]. The
difference among them is that the local version takes into account only the instances
that are in the same partition, and the global version considers all the instances in
the datasets. These two versions of the k-nearest neighbors imputation method are
analyzed in Sect. 6.5. In Fig. 9.6 we can find this algorithm available publicly as a
Spark Package.

```
SparkPackages                                    Smart_Imputation

This contribution implements two approaches of the k Nearest Neighbor Im-
putation focused on the scalability in order to handle big dataset. k Nearest
Neighbor - Local Imputation and k Nearest Neighbor Imputation - Global Im-
putation. The global proposal takes into account all the instances to calculate
the k nearest neighbors. The local proposal considers those that are into the
same partition, achieving higher times, but losing the information because it
does not consider all the samples.
- - - - - - - - - - - - - - - - - - - - - - - - - - - - - - - - - - - - - - - - - -
spark-shell --packages JMailloH:Smart_Imputation:1.0
- - - - - - - - - - - - - - - - - - - - - - - - - - - - - - - - - - - - - - - - - -
https://spark-packages.org/package/JMailloH/Smart_Imputation
```

Fig. 9.6 Spark package: Smart_Imputation

9.3.5 Discretization

The library also contains two distributed and parallel discretizers for dealing with
huge amounts of data: A Distributed Minimum Description Length Discretizer
(DMDLP) [33], and a Distributed Evolutionary Multivariate Discretizer (DEMD)
[31]. These two discretizers are described in Sects. 7.3 and 7.6, respectively. Both
of these algorithms are also available as Apache Spark packages.

- DMDLP is a distributed discretizer that implements Fayyad's discretizer [14].
  It is based on Minimum Description Length Principle for treating non discrete
  datasets from a distributed perspective. It supports sparse data, multi-attribute
  processing, and also is capable of dealing with attributes with a huge number of
  boundary points (<100K boundary points per attribute). In Fig. 9.7 we can find
  a Spark Package associated with this research.
- DEMD is an evolutionary discretizer. It uses binary chromosomes with a wrapper
  fitness function that optimizes the interval selection problem by compensating
  two factors: the simpleness of the solutions and the classification accuracy. In
  order to make DEMD able to cope with huge amounts of data, the evaluation
  phase has been distributed, splitting the set of chromosomes and the dataset
  into different partitions. Then a random cross-evaluation process is performed.
  In Fig. 9.8 we can find this algorithm available publicly in the third-party Apache
  Spark Repository.

SparkPackages   spark-MDLP-discretization

This method implements Fayyad's discretizer based on Minimum Description
Length Principle (MDLP) in order to treat non discrete datasets from a distributed perspective. It supports sparse data, parallel-processing of attributes,
etc.

```
spark-shell --packages sramirez:spark-MDLP- discretization:1.4.1
```

https://spark-packages.org/package/sramirez/
spark-MDLP-discretization

Fig. 9.7 Spark package: DMDLP discretizer

SparkPackages   spark-DEMD-discretizer

Here, a Distributed Evolutionary Multivariate Discretizer (DEMD) for data
reduction on Spark is presented. This evolutionary-based discretizer uses binary chromosome representation and a wrapper fitness function. The algorithm is aimed at optimizing the cut points selection problem by trading-off
two factors: simplicity of solutions and its classification accuracy. In order to
alleviate the complexity derived from the evolutionary process, the complete
evaluation phase has been fully parallelized. For this purpose, both the set
of chromosomes and instances are split into different partitions and a random
cross-evaluation process between them is performed.

```
spark-shell --packages sramirez:spark-DEMD-discretizer:1.0
```

https://spark-packages.org/package/sramirez/
spark-DEMD-discretizer

Fig. 9.8 Spark package: DEMD discretizer

9.3.6   Imbalanced Learning

Three popular methods for balancing a dataset are available in the library: Random
OverSampling (ROS), Random UnderSampling (RUS) [8], and SMOTE [11].

ROS reaches a balance in the data by replicating randomly instances from the
minority class from the original data, until the number of instances from both classes

is the same (or until a replication factor is reached). Depending on the posterior learning algorithm, the replication of instances may lead to overfitting.

On the other hand, RUS balances the dataset randomly removing instances from the majority class until the number of instances for both classes are identical. This approach works best when there is a high redundancy in the dataset, and achieves a lighter representation of the data storage-wide.

These two popular algorithms are analyzed in Sect. 8.2. In Fig. 9.9 we can find a Spark Package associated with ROS and RUS algorithms.

The library also includes an exact and fully scalable SMOTE in Spark for Big Data [7]. It is a Big Data implementation of the classic SMOTE algorithm. SMOTE forms new minority class examples by interpolating between several neighbor minority class examples. The SMOTE algorithm is analyzed in depth Sect. 8.3. In Fig. 9.10 we can find a Spark Package associated with the SMOTE algorithm.

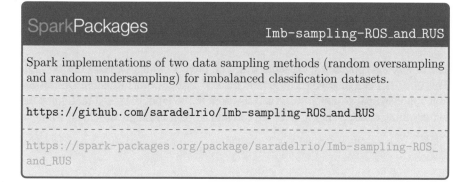

Fig. 9.9 Spark package: ROS-BD and RUS-BD

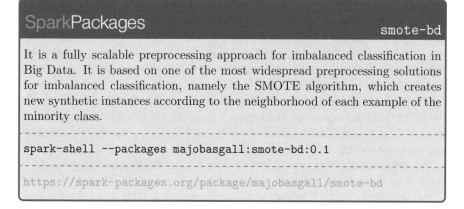

Fig. 9.10 Spark package: SMOTE-BD

SparkPackages   PCARD

This method implements the PCARD ensemble algorithm. PCARD ensemble method is a distributed upgrade of the method presented by A. Ahmad. The algorithm performs Random Discretization and Principal Components Analysis to the input data, then joins the results and trains a decision tree on it.

```
spark-shell --packages djgg:PCARD:1.3
```

https://spark-packages.org/package/djgg/PCARD

Fig. 9.11 Spark package: PCARD ensemble

9.3.7 Random Discretization and PCA Classifier

The library also contains a classifier based on preprocessing, named PCARD [21]. This classifier is a distributed ensemble method that performs Random Discretization and Principal Components Analysis, both to the input data, and then joins the two resulting datasets. In Fig. 9.11 we can find a Spark Package associated with this research.

9.4 FlinkML

FlinkML is the name of Apache Flink's [4] distributed ML library. Flink is a large-scale data processing framework, and it has at its core a dataflow engine. Like Apache Spark, it provides expressive, powerful, and easy to use APIs. Flink provides APIs for batch and stream data processing using a functional style syntax. This allows the developer to be familiar with the developing, and also to build prototypes fast.

Flink works in the Apache Hadoop Ecosystem [24]. At the core it has the streaming dataflow engine and, below that, it can have any kind of storage system. It also provides high availability. Similarly to Spark, Flink has a pipelining mechanism that allows to quickly build complex data analysis pipelines.

The core of Flink is the streaming dataflow engine that allows to deploy operators at the beginning of the program, and then pipe data through it. Using this kind of configuration enables Flink to do real-time streaming. This also allows to have iterations in Flink. In a batch processing system like Apache Spark, in order to run any iterative program, a new job must be scheduled at the start of each iteration, and that would create some overhead. Instead to do that, Flink has a native iterator support, which means that it can have partial solutions that will be updated frequently. This partial solutions, in the FlinkML case, would be the models, and it is able to iteratively update them in a streaming fashion. In addition to batch iterations, Flink also has support for what is called delta iterations. Delta iterations allow to shrink the size of the problem as it approaches closer to a solution.

As stated previously, the core feature of Flink is that it supports true streaming. On top of the data processing API, Flink has libraries that allows graph processing (Gelly), ML (FlinkML), SQL-like expression language (Table API), and complex event processing (FlinkCEP).

In this section we are focusing on FlinkML and its components. FlinkML is developed having three goals in mind:

1. Be really scalable: focusing on efficient algorithms and their implementations that really scale to web-scale data.
2. Minimize glue code: glue code can be defined as all the code necessary in a ML application in order to make it work. All the code that is necessary to deploy it. Google claims that mature applications contain 95% of glue code and 5% of actual logic [37]. This is what FlinkML tries to minimize, providing one system where everything can be performed.
3. Ease of use: provide easy to use and intuitive APIs, and also focus on creating a good documentation and examples of use.

Here, we describe and classify all DM techniques for Flink's *FlinkML* machine learning library.[2]

9.4.1 Data Preprocessing

Three simple yet very popular and widely used data preprocessing algorithms are included in FlinkML. Here we outline the three different algorithms:

- Polynomial Features: it generates a vector containing all the polynomial combinations with degree less than or equal to a given degree. Flink's polynomial features implementation orders the these polynomials in descending order according to their degree.
- StandardScaler: scales the data to have zero mean and a standard deviation equal to 1. The user can specify its own desired mean and standard deviation.
- MinMaxScaler: scales the data to a specific range [min, max]. By default all values will be scaled to [0, 1] interval.

---

[2]*FlinkML* documentation: https://ci.apache.org/projects/flink/flink-docs-master/dev/libs/ml/.

9.4.2   Recommendation

FlinkML also includes a popular recommendation algorithm, Alternating Least Squares (ALS) [40]. This algorithm factorizes a given matrix R into two factors U and V such that $R \approx U^T V$.

Flink implementation includes a regularization scheme in order to avoid overfitting, namely weighted-λ-regularization [51].

9.4.3   Outlier Selection

FlinkML implements Stochastic Outlier Selection (SOS) [25] for outlier detection. An outlier can be defined as an example that is deviated enough from the majority of the data points.

SOS is an unsupervised outlier-selection algorithm. It applies affinity-based outlier selection to a set of vectors, and outputs a probability of being an outlier for each example. A data point can be considered an outlier when the rest of the data points have not enough affinity with it.

9.4.4   Utilities

Two additional utilities are provided for distance calculation and validation purposes:

- Distance Metrics: different distance metrics are provided. The included distances are: Euclidean distance, Squared Euclidean distance, Cosine Similarity, Chebyshev Distance, Manhattan Distance, Minkowski Distance, and Tanimoto Distance.
- Cross Validation: in order to prevent overfitting, a validation strategy is recommended. FlinkML includes four different types of validation schemes: Train-Test splits, Train-Test-Holdout Splits, K-Fold Splits, and Multi-Random Splits.

9.5   BigDaPFlink

BigDaPFlink is a Big Data library oriented to online data preprocessing for Apache Flink [19]. This library contains six of the most popular and widely used algorithms for data preprocessing in data streaming [2]. It is composed of three feature selection algorithms and three discretization algorithms. In Fig. 9.12 we can find the algorithm implementations associated with this library.

SCI2S   BigDaPFlink

Big Data library oriented to online data preprocessing for Apache Flink. This
library contains six of the most popular and widely used algorithms for data
preprocessing in data streaming. It is composed of three feature selection
algorithms and three discretization algorithms.

https://sci2s.ugr.es/BigDaPFlink

Fig. 9.12 BigDaPFlink library

9.5.1 Feature Selection

The library contains three of the most popular feature selection algorithms for data
streaming in the literature: Information Gain, Online Feature Selection (OFS), and
Fast Correlation-Based Filter (FCBF).

- Information gain is a feature selection algorithm composed of two steps, an
  incremental feature ranking method, and an incremental learning algorithm that
  can consider a subset of the features during prediction (Naïve Bayes) [26].
  First, the conditional entropy with respect to the class is computed. Then, the
  information gain is calculated for each attribute. Finally, once the algorithm has
  all the information gains for each feature, it selects the best N as features.
- OFS is an ε-greedy online feature selection method based on feature weights
  generated by an online classifier (in this case a neural network) which makes a
  trade-off between exploration and exploitation of features [45].
- FCBF is a feature selection algorithm where the class relevance and the correlation between each feature pair of features are taken into account [49]. It is based
  on information theory, it uses Symmetrical Uncertainty to calculate dependencies
  of features and the class importance. It starts with the full set of features and,
  using a backward selection technique with a sequential search strategy, it removes
  all the irrelevant and redundant features. Finally, it stops when no more features
  are left to eliminate.

9.5.2 Discretization

In this section we show the three online discretization algorithms for data streaming
available in the library: Incremental Discretization Algorithm (IDA), Partition
Incremental Discretization Algorithm (PiD), and Local Online Fusion Discretizer

(LOFD). Discretization in data streaming has the challenge of the concept drift. These three methods tackle it in three different ways.

- IDA performs an approximate quantile-based discretization on the entire encountered data stream to date by keeping a random sample of the data [46]. This sample is then used to calculate the cut points of the dataset. It uses the reservoir sampling algorithm to maintain this sample randomly updated from the entire stream. In IDA a sample of the data is used because it is not feasible nor possible to keep the complete data stream in memory.
- PiD discretizes data streams in an incremental manner [17]. The discretization process is performed in two steps. The first step discretizes the data using more intervals than required, keeping some statistics of it. The second and final step is to use that statistics to create the final discretization. It is constant in time and space even for infinite streams, as PiD processes all the streaming examples in a single scan.
- LOFD is a very recent proposal for online data streaming discretization. It is an online and self-adaptive discretizer [32]. LOFD is capable of smoothly adapt its interval limits, reducing the negative impact of shifts (concept drift), and also to analyze the interval labeling and interaction problems. The interaction between the discretizer and the learner algorithm is addressed by providing two alike solutions. LOFD generates an online and self-adaptive discretization for streaming classification whose objective is to reduce the negative impact of fluctuations in evolving intervals.

9.6 Summary and Conclusions

In this chapter we have studied in depth the two most popular Big Data ML libraries for both batch and stream data processing. For static data, MLlib is the most popular and widely used ML library. It is built on top of Apache Spark and provides a useful set of algorithms and utilities. We have also presented a new Big Data library focused on data preprocessing for Apache Spark, BigDaPSpark. This library is composed of several state-of-the-art algorithms for data preprocessing in Big Data.

From the streaming point of view, FlinkML is the ML library of Apache Flink. It provides some algorithms for supervised and unsupervised learning, as well as some utilities, like cross validation or distance metrics. Additionally, we have introduced BigDaPFlink, a Big Data streaming library focused on streaming data preprocessing under Apache Flink. This library contains six data preprocessing algorithms for Big Data online scenarios.

BigDaPSpark and BigDaPFlink libraries are part of BigDaPTOOLS library. This library is composed of three libraries for the most popular and widely used ML and DM frameworks (R, Apache Spark, and Apache Flink). BigDaPTOOLS contains a wide variety of state-of-the-art data preprocessing algorithms. In Fig. 9.13 we can find the algorithm implementations associated with this library.

BigDaPTOOLS: Big Data Preprocessing: Models and Tools to improve the quality of data.
It is a Big Data preprocessing library, composed of three sub-libraries, for the most popular and widely used machine learning and data mining frameworks: R, Apache Spark and Apache Flink. It contains data preprocessing algorithms for many different disciplines, including imperfect data approaches, imbalance dataset preprocessing techniques, and data reduction algorithms. It also includes data preprocessing proposals for non-standard classification, including multilabel and data streaming problems.

https://sci2s.ugr.es/BigDaPTOOLS

Fig. 9.13 BigDaPTOOLS library

As we have seen in this chapter, new libraries for Big Data preprocessing are emerging in the last years. We can find an increasing list of data preprocessing algorithms ready to tackle Big Data problems. Apache Spark is positioning as the best performing and most complete Big Data framework for ML and DM. On the other hand, Apache Flink is becoming the reference in Big Data streaming scenarios. MLlib is growing at a faster pace than FlinkML. For this reason, it is desirable to develop new data preprocessing methods on Apache Flink, adapted to the Big Data streaming environment.

References

1. Al-Fuqaha, A., Guizani, M., Mohammadi, M., Aledhari, M., & Ayyash, M. (2015). Internet of things: A survey on enabling technologies, protocols, and applications. *IEEE Communications Surveys Tutorials, 17*(4), 2347–2376.
2. Alcalde-Barros, A., García-Gil, D., García, S., & Herrera, F. (2019). DPASF: A Flink library for streaming data preprocessing. *Big Data Analytics, 4*(1), 4.
3. Angiulli, F. (2007). Fast nearest neighbor condensation for large data sets classification. *IEEE Transactions on Knowledge and Data Engineering, 19*(11), 1450–1464.
4. Apache Flink. (2019). Apache Flink. http://flink.apache.org/.
5. Armbrust, M., Xin, R. S., Lian, C., Huai, Y., Liu, D., Bradley, J. K., et al. (2015). Spark SQL: Relational data processing in spark. In *ACM SIGMOD International Conference on Management of Data, SIGMOD '15* (pp. 1383–1394).
6. Arnaiz-González, Á., González-Rogel, A., Díez-Pastor, J.-F., & López-Nozal, C. (2017). MR-DIS: democratic instance selection for big data by MapReduce. *Progress in Artificial Intelligence, 6*(3), 211–219.

7. Basgall, M. J., Hasperué, W., Naiouf, M., Fernández, A., & Herrera, F. (2018). SMOTE-BD: An exact and scalable oversampling method for imbalanced classification in big data. *Journal of Computer Science and Technology, 18*(03), e23.
8. Batista, G. E. A. P. A., Prati, R. C., & Monard, M. C. (2004). A study of the behavior of several methods for balancing machine learning training data. *SIGKDD Explorations Newsletter, 6*(1), 20–29.
9. Blum, A. L., & Langley, P. (1997). Selection of relevant features and examples in machine learning. *Artificial Intelligence, 97*(1–2), 245–271.
10. Buitinck, L., Louppe, G., Blondel, M., Pedregosa, F., Mueller, A., Grisel, O., et al. (2013). API design for machine learning software: experiences from the scikit-learn project. In *ECML PKDD Workshop: Languages for Data Mining and Machine Learning* (pp. 108–122).
11. Chawla, N. V., Bowyer, K. W., Hall, L. O., & Kegelmeyer, W. P. (2002). SMOTE: Synthetic minority over-sampling technique. *Journal of Artificial Intelligence Research, 16*, 321–357.
12. Chen, H., Chiang, R. H. L., & Storey, V. C. (2012). Business intelligence and analytics: From big data to big impact. *MIS Quarterly, 36*(4), 1165–1188.
13. Dean, J., & Ghemawat, S. (2010). MapReduce: A flexible data processing tool. *Communications of the ACM, 53*(1), 72–77.
14. Fayyad, U. M., & Irani, K. B. (1993). Multi-interval discretization of continuous-valued attributes for classification learning. In *IJCAI* (pp. 1022–1029).
15. Fernández, A., del Río, S., López, V., Bawakid, A., del Jesús, M. J., Benítez, J. M., et al. (2014). Big data with cloud computing: an insight on the computing environment, MapReduce, and programming frameworks. *Wiley Interdisciplinary Reviews: Data Mining and Knowledge Discovery, 4*(5), 380–409.
16. Figueredo, G. P., Triguero, I., Mesgarpour, M., Guerra, A. M., Garibaldi, J. M., & John, R. I. (2017). An immune-inspired technique to identify heavy goods vehicles incident hot spots. *IEEE Transactions on Emerging Topics in Computational Intelligence, 1*(4), 248–258.
17. Gama, J., & Pinto, C. (2006). Discretization from data streams: Applications to histograms and data mining. In *Proceedings of the 2006 ACM Symposium on Applied Computing* (pp. 662–667). New York: ACM.
18. García, S., Cano, J. R., & Herrera, F. (2008). A memetic algorithm for evolutionary prototype selection: A scaling up approach. *Pattern Recognition, 41*(8), 2693–2709.
19. García-Gil, D., Alcalde-Barros, A., Luengo, J., García, S., & Herrera, F. (2019). Big data preprocessing as the bridge between big data and smart data: BigDaPSpark and BigDaPFlink libraries. In *Proceedings of the 4th International Conference on Internet of Things, Big Data and Security - Volume 1: IoTBDS* (pp. 324–331). INSTICC, SciTePress.
20. García-Gil, D., Luengo, J., García, S., & Herrera, F. (2019). Enabling smart data: Noise filtering in big data classification. *Information Sciences, 479*, 135–152.
21. García-Gil, D., Ramírez-Gallego, S., García, S., & Herrera, F. (2018). Principal components analysis random discretization ensemble for big data. *Knowledge-Based Systems, 150*, 166–174.
22. Gupta, P., Sharma, A., & Jindal, R. (2016). Scalable machine learning algorithms for big data analytics: A comprehensive review. *Wiley Interdisciplinary Reviews: Data Mining and Knowledge Discovery, 6*(6), 194–214.
23. Guyon, I., Gunn, S., Nikravesh, M., & Zadeh, L. A. (2006). *Feature extraction: Foundations and applications (Studies in fuzziness and soft computing)*. New York: Springer.
24. Hadoop Distributed File System. (2019). Hadoop Distributed File System. https://hadoop.apache.org/docs/stable/hadoop-project-dist/hadoop-hdfs/HdfsUserGuide.html.
25. Janssens, J., Huszár, F., Postma, E. O., & van den Herik, H. J. (2012). Stochastic outlier selection. Technical Report, Technical report TiCC TR 2012–001, Tilburg University.
26. Katakis, I., Tsoumakas, G., & Vlahavas, I. (2005). On the utility of incremental feature selection for the classification of textual data streams. In *Panhellenic Conference on Informatics* (pp. 338–348). Berlin: Springer.
27. Marx, V. (2013). Biology: The big challenges of big data. *Nature, 498*(7453), 255–260.

28. Meng, X., Bradley, J., Yavuz, B., Sparks, E., Venkataraman, S., Liu, D., et al. (2016). Mllib: Machine learning in Apache spark. *Journal of Machine Learning Research, 17*(34), 1–7.
29. Philip-Chen, C. L., & Zhang, C. Y. (2014). Data-intensive applications, challenges, techniques and technologies: A survey on big data. *Information Sciences, 275*(10), 314–347.
30. Ramírez-Gallego, S., Fernández, A., García, S., Chen, M., & Herrera, F. (2018). Big data: Tutorial and guidelines on information and process fusion for analytics algorithms with MapReduce. *Information Fusion, 42*, 51–61.
31. Ramírez-Gallego, S., García, S., Benítez, J. M., & Herrera, F. (2018). A distributed evolutionary multivariate discretizer for big data processing on Apache spark. *Swarm and Evolutionary Computation, 38*, 240–250.
32. Ramírez-Gallego, S., García, S., & Herrera, F. (2018). Online entropy-based discretization for data streaming classification. *Future Generation Computer Systems, 86*, 59–70.
33. Ramírez-Gallego, S., García, S., Mouriño-Talín, H., Martínez-Rego, D., Bolón-Canedo, V., Alonso-Betanzos, A., et al. (2016). Data discretization: Taxonomy and big data challenge. *Wiley Interdisciplinary Reviews: Data Mining and Knowledge Discovery, 6*(1), 5–21.
34. Ramírez-Gallego, S., Mouriño-Talín, H., Martínez-Rego, D., Bolón-Canedo, V., Benítez, J. M., Alonso-Betanzos, A., et al. (2018). An information theory-based feature selection framework for big data under Apache spark. *IEEE Transactions on Systems, Man, and Cybernetics: Systems, 48*(9), 1441–1453.
35. Sánchez, J. S., Barandela, R., Marqués, A. I., Alejo, R., & Badenas, J. (2003). Analysis of new techniques to obtain quality training sets. *Pattern Recognition Letters, 24*(7), 1015–1022.
36. Sánchez, J. S., Pla, F., & Ferri, F. J. (1997). Prototype selection for the nearest neighbour rule through proximity graphs. *Pattern Recognition Letters, 18*(6), 507–513.
37. Sculley, D., Holt, G., Golovin, D., Davydov, E., Phillips, T., Ebner, D., et al. (2014). Machine learning: The high interest credit card of technical debt. In *SE4ML: Software Engineering for Machine Learning (NIPS 2014 Workshop)*.
38. Skalak, D. B. (1994). Prototype and feature selection by sampling and random mutation hill climbing algorithms. In *Machine Learning Proceedings 1994* (pp. 293–301). Amsterdam: Elsevier.
39. Snir, M., & Otto, S. (1998). *MPI-The complete reference: The MPI core*. Cambridge, MA: MIT Press.
40. Takane, Y., Young, F. W., & De Leeuw, J. (1977). Nonmetric individual differences multidimensional scaling: An alternating least squares method with optimal scaling features. *Psychometrika, 42*(1), 7–67.
41. Tomek, I. (1976). An experiment with the edited nearest-neighbor rule. *IEEE Transactions on systems, Man, and Cybernetics, SMC-6*(6), 448–452.
42. Triguero, I., García, S., & Herrera, F. (2011). Differential evolution for optimizing the positioning of prototypes in nearest neighbor classification. *Pattern Recognition, 44*(4), 901–916.
43. Triguero, I., García-Gil, D., Maillo, J., Luengo, J., García, S., & Herrera, F. (2019). Transforming big data into smart data: An insight on the use of the k-nearest neighbors algorithm to obtain quality data. *Wiley Interdisciplinary Reviews: Data Mining and Knowledge Discovery, 9*(2), e1289.
44. Triguero, I., Peralta, D., Bacardit, J., García, S., & Herrera, F. (2015). MRPR: A MapReduce solution for prototype reduction in big data classification. *Neurocomputing, 150*, 331–345.
45. Wang, J., Zhao, P., Hoi, S. C. H., & Jin, R. (2014). Online feature selection and its applications. *IEEE Transactions on Knowledge and Data Engineering, 26*(3), 698–710.
46. Webb, G. I. (2014). Contrary to popular belief incremental discretization can be sound, computationally efficient and extremely useful for streaming data. In *2014 IEEE International Conference on Data Mining* (pp. 1031–1036).
47. White, T. (2012). *Hadoop: The definitive guide* (3rd ed.). Sebastopol, CA: O'Reilly Media.
48. Wilson, D. L. (1972) Asymptotic properties of nearest neighbor rules using edited data. *IEEE Transactions on Systems, Man, and Cybernetics, SMC-2*(3), 408–421.

49. Yu, L., & Liu, H. (2003). Feature selection for high-dimensional data: A fast correlation-based filter solution. In *Proceedings of the 20th International Conference on Machine Learning (ICML-03)* (pp. 856–863).
50. Zaharia, M., Chowdhury, M., Das, T., Dave, A., Ma, J., McCauley, M., et al. (2012). Resilient distributed datasets: A fault-tolerant abstraction for in-memory cluster computing. In *Proceedings of the 9th USENIX Conference on Networked Systems Design and Implementation* (pp. 1–14).
51. Zhou, Y., Wilkinson, D., Schreiber, R., & Pan, R. (2008). Large-scale parallel collaborative filtering for the Netflix prize. In R. Fleischer & J. Xu (Eds.), *Algorithmic aspects in information and management* (pp. 337–348). Berlin/Heidelberg: Springer.

Chapter 10
Final Thoughts: From Big Data to Smart Data

In previous chapters we have discussed both the advantages and necessity of having a correct data treatment methodology. As we have seen, such a treatment is necessary for several reasons: data quality enhancement, faster learning times, less space requirements, and Smart Data generation. A well-known principle in computer science is the principle of *garbage in–garbage out*: no matter how good your model is created, as long as the data is bad, the results will be poor (Fig. 10.1). Thus, obtaining quality, Smart Data must be first data scientist's objective. The way to obtain such a Smart Data is to apply data preprocessing.

The plethora of different preprocessing techniques and the constant proposal of algorithms in the specialized literature poses a challenge to any practitioner [2]. Having a unique preprocessing methodology for all datasets is unfeasible, as the different combination of preprocessing steps will yield different results [3]. The presence of Big Data does not ease this issue: large volumes of data lead to the appearance of new problems, as data redundancy, which may render useless traditional preprocessing techniques. Missing values imputation, for instance, cannot be feasible compared to eliminating incomplete instances, as data redundancy will provide enough examples to avoid bias generated by such an elimination. Therefore, classical approaches must be revised before tackling Big Data preprocessing.

We cannot forget the importance that Deep Learning has nowadays [5, 6]. While Deep Learning also needs large collections of data in order to obtain generalizable and accurate model, the nature of such data is different of that used by Big Data frameworks. While the latter is oriented to textual, tuple-based data, Deep Learning is successful thanks to its ability to deal with non-standard data types: images, voice, sound, etc. Deep Learning models are robust and with enough data they may yield accurate solutions to almost every problem [10]. However, the lack of data for a given problem might force practitioners to elect for data augmentation techniques [1]. Data augmentation enables the user to increase the amount of training examples from a reduced data pool size, but transforming the training data to generate synthetic examples may lead to noise introduction if not

© Springer Nature Switzerland AG 2020
J. Luengo et al., *Big Data Preprocessing*,
https://doi.org/10.1007/978-3-030-39105-8_10

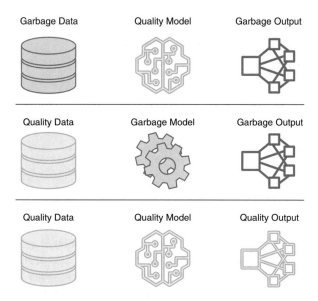

Fig. 10.1 Garbage in, garbage out principle

performed correctly [7, 9]. This kind of "noisy examples" and their treatment is an open challenge that may constitute a bridge between Deep Learning and Big Data preprocessing, due to the large amount of examples that can be found in both paradigms. However, this complex issue is out of the scope of this book but still an interesting challenge.

Data preprocessing in data mining for business has been known to be very costly, both in time and computational resources [11]. Actually, datasets collection and preprocessing accounts for about 80% of the work of data scientists, according to recent surveys as shown in Fig. 10.2 (source: Forbes 2016[1]). These costs are even greater in Big Data environments, where the computationally expensive techniques need to deal with massive data. The growing attention to Smart Data in the companies [4] is boosting even more the share hold by preprocessing in the whole knowledge extraction process, where cleaning and preparing the data is overtaking data integration as the most devoted task.

Direct translation of sequential algorithms is unfeasible, requiring novel approaches and designs to cope with the exponential data growth rates [8]. The challenge imposed by the new designs has limited the rate of Big Data preprocessing proposals, which still constitutes a prominent field of work with many unresolved problems that need to be tackled. There is still a long journey to go for data scientists in Big Data preprocessing.

---

[1]https://www.forbes.com/sites/gilpress/2016/03/23/data-preparation-most-time-consuming-least-enjoyable-data-science-task-survey-says/#5a932e1b6f63.

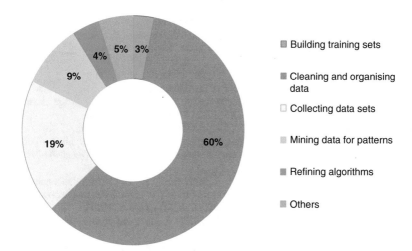

Fig. 10.2 Data scientists spend 60% of their time on cleaning and organizing data. Collecting datasets comes second at 19% of their time, meaning data scientists spend around 80% of their time on preparing and managing data for analysis

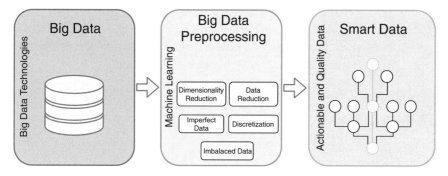

Fig. 10.3 Big Data preprocessing is the key to transform raw Big Data into quality Smart Data

We hope the book has served the reader as an appropriate introduction to the Big Data world and how to reach Smart Data, where Big Data preprocessing is the bridge between both sides of the data river towards quality data, as depicted in Fig. 10.3.

References

1. Castillo, A., Tabik, S., Pérez, F., Olmos, R., & Herrera, F. (2019). Brightness guided preprocessing for automatic cold steel weapon detection in surveillance videos with deep learning. *Neurocomputing, 330*, 151–161.
2. García, S., Luengo, J., & Herrera, F. (2015). *Data preprocessing in data mining*. Berlin: Springer.

3. García, S., Luengo, J., & Herrera, F. (2016). Tutorial on practical tips of the most influential data preprocessing algorithms in data mining. *Knowledge-Based Systems, 98*, 1–29.
4. George, G., Haas, M. R., & Pentland, A. (2014). Big data and management. *Academy of Management Journal, 57*(2), 321–326.
5. Goodfellow, I., Bengio, Y., & Courville, A. (2016). *Deep learning*. Cambridge: MIT Press.
6. LeCun, Y., Bengio, Y., & Hinton, G. (2015). Deep learning. *Nature, 521*(7553), 436.
7. Perez, L., & Wang, J. (2017). The effectiveness of data augmentation in image classification using deep learning. arXiv preprint. arXiv:1712.04621.
8. Triguero, I., García-Gil, D., Maillo, J., Luengo, J., García, S., & Herrera, F. (2019). Transforming big data into smart data: An insight on the use of the k-nearest neighbors algorithm to obtain quality data. *Wiley Interdisciplinary Reviews: Data Mining and Knowledge Discovery, 9*(2), e1289.
9. Wong, S. C., Gatt, A., Stamatescu, V., & McDonnell, M. D. (2016). Understanding data augmentation for classification: when to warp? In *2016 international conference on digital image computing: techniques and applications (DICTA)* (pp. 1–6). IEEE.
10. Zhang, Q., Yang, L. T., Chen, Z., & Li, P. (2018). A survey on deep learning for big data. *Information Fusion, 42*, 146–157.
11. Zhang, S., Zhang, C., & Yang, Q. (2003). Data preparation for data mining. *Applied Artificial Intelligence, 17*(5–6), 375–381.

Printed in the United States
by Baker & Taylor Publisher Services